SEMIGROUPS IN GEOMETRICAL FUNCTION THEORY

Semigroups in Geometrical Function Theory

by

David Shoikhet

Department of Mathematics,
Technion-Israel Institute of Technology,
Haifa, Israel

KLUWER ACADEMIC PUBLISHERS
DORDRECHT / BOSTON / LONDON

A C.I.P. Catalogue record for this book is available from the Library of Congress.

ISBN 978-90-481-5747-1

Published by Kluwer Academic Publishers,
P.O. Box 17, 3300 AA Dordrecht, The Netherlands.

Sold and distributed in North, Central and South America
by Kluwer Academic Publishers,
101 Philip Drive, Norwell, MA 02061, U.S.A.

In all other countries, sold and distributed
by Kluwer Academic Publishers,
P.O. Box 322, 3300 AH Dordrecht, The Netherlands.

Printed on acid-free paper

Contents

Preface   vii

Preliminaries   **1**
    0.1   Notations and notions   1
    0.2   Holomorphic functions of a complex variable   4
    0.3   Convergence of holomorphic functions   6
    0.4   Metric spaces and fixed point principles   7

1   The Wolff–Denjoy theory on the unit disk   **9**
    1.1   Schwarz–Pick Lemma and automorphisms   9
    1.2   Boundary behavior of holomorphic self-mappings   17
    1.3   Fixed points of holomorphic self-mappings   25
    1.4   Fixed point free holomorphic self-mappings of Δ.
        The Denjoy–Wolff Theorem.   32
    1.5   Commuting family of holomorphic mappings of the unit disk. . . .   36

2   Hyperbolic geometry on the unit disk and fixed points   **39**
    2.1   The Poincaré metric on Δ   39
    2.2   Infinitesimal Poincaré metric and geodesics   44
    2.3   Compatibility of the Poincaré metric with convexity   46
    2.4   Fixed points of ρ-nonexpansive mappings on the unit disk   52

3   Generation theory on the unit disk   **59**
    3.1   One-parameter continuous semigroup of holomorphic and
        ρ-nonexpansive self-mappings   59
    3.2   Infinitesimal generator of a one-parameter continuous semigroup .   62
    3.3   Nonlinear resolvent and the exponential formula   67
    3.4   Monotonicity with respect to the hyperbolic metric   79
    3.5   Flow invariance conditions for holomorphic functions   83
    3.6   The Berkson–Porta parametric representation of semi-complete
        vector fields   95

4   Asymptotic behavior of continuous flows   **101**
    4.1   Stationary points of a flow on Δ   101
    4.2   Null points of complete vector fields   104

v

4.3   Embedding of discrete time group into a continuous flow 109

4.4   Rates of convergence of a flow with an interior stationary point . . 113

4.5   A rate of convergence in terms of the Poincaré metric 120

4.6   Continuous version of the Julia–Wolff–Carathéodory Theorem . . . 124

4.7   Lower bounds for ρ-monotone functions 135

5 Dynamical approach to starlike and spirallike functions   153

5.1   Generators on biholomorphically equivalent domains 154

5.2   Starlike and spirallike functions 157

5.3   A generalized Visser–Ostrowski condition and fanlike functions . . 163

5.4   An invariance property and approximation problems 166

5.5   Hummel's multiplier and parametric representations of
starlike functions 172

5.6   A conjecture of Robertson and geometrical characteristics of
fanlike functions 176

5.7   Converse theorems on starlike, spirallike and fanlike functions . . . 186

5.8   Growth estimates for spirallike, starlike and fanlike functions . . . 194

5.9   Remarks on Schroeder's equation and the Koenigs
embedding property 198

Bibliography   205

Author and Subject Index   216

List of figures   221

Preface

Historically, complex analysis and geometrical function theory have been intensively developed from the beginning of the twentieth century. They provide the foundations for broad areas of mathematics. In the last fifty years the theory of holomorphic mappings on complex spaces has been studied by many mathematicians with many applications to nonlinear analysis, functional analysis, differential equations, classical and quantum mechanics. The laws of dynamics are usually presented as equations of motion which are written in the abstract form of a dynamical system: $dx/dt + f(x) = 0$, where x is a variable describing the state of the system under study, and f is a vector function of x. The study of such systems when f is a monotone or an accretive (generally nonlinear) operator on the underlying space has been recently the subject of much research by analysts working on quite a variety of interesting topics, including boundary value problems, integral equations and evolution problems (see, for example, [19, 13] and [29]).

In a parallel development (and even earlier) the generation theory of one-parameter semigroups of holomorphic mappings in $\mathbb{C}^n$ has been the topic of interest in the theory of Markov stochastic processes and, in particular, in the theory of branching processes (see, for example, [63, 127, 48] and [69]). Later, such semigroups appeared in other fields: one-dimensional complex analysis ([17, 28]), finite-dimensional manifolds [2, 5], Banach spaces geometry [12, 142], control and optimization theory [64], and Krein spaces [147, 148].

At the same time very rapid developments in multi-dimensional complex analysis, functional analysis and a variety of techniques and methods have diverted attention from the roots of one-dimensional dynamic approaches. Furthermore, some interesting results on evolution equations, presented in various papers in journals on nonlinear analysis, abstract analysis, and differential equations, have remained unnoticed by experts in one-dimensional complex variable.

One of the first applicable models of the complex dynamical systems on the unit disk arose more than a hundred years ago from studies of the dynamics of stochastic branching processes.

In 1874 F. Galton and H.W. Watson [151] in treating the problem of the extinction probability of family names, formulated a mathematical model in terms of the probability generating function:

$$F(z) = \sum_{k=0}^{\infty} p_k z^k, \quad |z| \le 1,$$

where z is a complex variable, $p_0, p_1, \ldots, p_k$ are nonnegative numbers (probabilities) such that $\sum_{k=0}^{\infty} p_k = 1$, and its iterations:

$$F^{(0)}(z) = z, \quad F^{(n+1)}(z) = F^{(n)}(F(z)).$$

The first complete and correct determination of the extinction probability for the Galton–Watson process as the limit points of the iteration sequence was given by J.F. Steffensen in 1930 [139]. Since then the interest in this model has increased because of connections with chemical and nuclear chain reactions, the study of the multiplication of electrons in the electron multiplier, the theory of cosmic radiation, and many other biological and physical problems.

Detailed description of classical results concerning branching processes can be found in the books of T.E. Harris [63] and B.A. Sevastyanov [127]. We only note here that while the original Galton–Watson process has been related to discrete time branching process (i.e., it is described by an iteration process of a single probability generating function) the further development involved also the consideration of continuous time branching processes based on one-parameter semigroups of analytical self-mappings of the unit disk. One of the problems in analysis is, given a function $F(z)$ find a function $F(z, t)$, with $F(z, 1) = F(z)$ satisfying the semigroup property

$$F(z, t + s) = F(F(t, z), s), \quad t, s \ge 0,$$

where z is a complex variable. Since this formula expresses the characteristic property of iteration when t and s are integers we may consider $F(z, t)$ as a fractional iterate of F, when t is not an integer.

G. Koenigs (1884) [78] showed how this problem may be solved, if F is analytic self-mapping on the unit disk with an interior fixed point $z_0 = F(z_0)$, such that $0 < |F'(z_0)| < 1$, by considering the convergence of the sequence $\{F^{(n)}(z)\}$ to z_0, as $n \to \infty$ in a neighborhood of the point z_0.

These and other problems led to the following general question:

Let D be a bounded domain in $\mathbb{C}$ and let $F : D \mapsto D$ be a holomorphic mapping of D into itself. Does the sequence of iterates $\{F^{(n)}(z)\}_{n=1}^{\infty}$ converge uniformly on compact subsets of D to some holomorphic mapping $h : D \mapsto \mathbb{C}$?

In 1926 J. Wolff (see [155]–[157]) and A. Denjoy [31] solved this problem for $D = \Delta$, the unit disk in $\mathbb{C}$. Applying Schwarz's and Julia's Lemmas they proved the following remarkable results:

Let $F : \Delta \mapsto \Delta$ be a holomorphic mapping of the unit disk of $\mathbb{C}$ into itself.

If F is not an automorphism of Δ, with exactly one interior fixed point, then $\{F^{(n)}(z)\}$ converges uniformly on compact subsets of Δ to a holomorphic mapping $h : \Delta \mapsto \bar{\Delta}$.

Moreover, if F is not the identity then h is a constant.

This result has given a powerful thrust to the development of different aspects of complex dynamical systems on the unit disk, the complex plane, or, more generally, hyperbolic Riemann surfaces.

In 1930 G. Julia [72] in publishing of geometrical principles of analysis characterized the dynamics of analytic motions in the unit disk. Over the last 40 years these results have been developed in at least three directions.

In 1960 V.P. Potapov [110] extended the classical lemma of G. Julia to matrix-valued holomorphic mappings of a complex variable. Next I. Glicksberg [49] and K. Fan [43] established the version of Julia's lemma for function algebras. Also K. Fan [44, 45] extended a result of J. Wolff [155] on iterates of self-mapping of proper contraction operator in the sense of functional calculus.

Recently K. Wlodarczyk [152]–[154] and P. Mellon [94] motivated by the research of V. Potapov and K. Fan generalized these results to holomorphic mappings on J^*-algebras. In general, they used operator theoretic methods.

Finite-dimensional generalizations are to be found in G.N. Chen [24], Y. Kubota [80], B. MacCluer [91], M. Abate [2, 5], P. Mercer [96], among others.

In all cases it appears that some sort of 'finiteness' or 'compactness' is required. E. Vesentini [144, 145], P. Mazet and J.P. Vigué [92, 93] have used spectral analysis for the case when a holomorphic self-mapping of a domain in a Banach space has an interior fixed point.

Further observations of the Wolff–Denjoy Theory (even for one-dimensional cases) yield extensions to those mappings which are not necessarily holomorphic but are nonexpansive with respect to a hyperbolic metric on a domain.

This approach has been developed by C. Earle and R. Hamilton [34], K. Goebel, T. Sekowski, T. Kuczumov, A. Stachura, S. Reich, I. Shafrir including others (see, for example, [50, 52, 53, 52, 82, 83, 129, 130]) and it is based on the study of the so called approximating curves.

These results can also be considered as an implicit analogues of the Denjoy–Wolff Theorem. It is also remarkable that the asymptotic behavior of the approximating curves is actually nicer than that of the usual iterative process. Moreover, these implicit methods may be useful not only for a self-mapping of a domain, but also for a wider class of mappings which satisfy certain flow invariance conditions.

Actually, the study of these methods shows their deep connections with continuous time dynamical systems.

In fact, in 1964 F. Forelli [46] established a one-to-one correspondence of the group of linear isometrics of a Hardy space H^p, $p > 0$, $p \neq 1$, and the group of Möbius transformations on the unit disk. E. Berkson and H. Porta in [17] established a continuous analog of the Denjoy–Wolff Theorem for continuous semigroups of holomorphic self-mappings of the unit disk. This approach has been used by several mathematicians to study the asymptotic behavior of solutions of Cauchy problems (see, for example, [16, 2, 5, 28] and [117]).

Berkson and Porta [17] also apply their continuous analogue of the Denjoy–Wolff Theorem to the study of the eigenvalue problem of composition operators in Hardy spaces. Similar approaches were used by A.G. Siskakis [136] to study the Césaro operator on H^p [136], F. Jafari and K. Yale [70], Y. Latuskin and M. Stepin [85] for weighted composition operators and dynamical systems on H^p. See also the book of C.C. Cowen and B. MacCluer [28] and references there.

These arguments motivate us to give a systematic exposition of the Wolff–Denjoy theory and its application to dynamical systems. In fact, it does not seem possible to cover the extensive literature concerning this subject in a single book. Nevertheless, I believe that the first step in this direction should be the understanding of dynamic processes on the unit disk of the complex plane. Actually, one can see that multi-dimensional generalizations may often be obtained by the reduction to the one-dimensional case, or by generalizations of attendant notions to higher dimensions.

The study of images of domains under holomorphic (or biholomorphic) transformations links some of the most basic questions one can ask about semigroups with nice classical results in geometric function theory. As a part of analytic function theory, research on the geometry of domains in the complex plane is of old origin, dating back to the 19th century. The basic methods of geometrical function theory include the square principles, methods of contour integration, variational methods, extreme metrics, and integral representation theory. Good introductions in these topics can be found in the books [55, 87, 11] and [57]. An exhaustive bibliography on geometrical function theory was compiled by S.D. Bernardi [18] before and until 1981.

This book is not intended to be a survey of the theories mentioned above. Our primary focus will be only on the material investing old theorems with new meanings which are related to applications of evolution equations to the geometry of domains in the complex plane. For example, it is a well known result, due to R. Nevanlinna (1921) [102], that *if h is holomorphic in $|z| < 1$ and satisfies $h(0) = 0$, $h'(0) \neq 0$, then h is univalent and maps the unit disk onto a starshaped domain (with respect to 0) if and only if $\mathrm{Re}[zh'(z)/h(z)] > 0$ everywhere.* This result, as well as most of the work on starlike functions on the unit disk, can be obtained from the identity

$$\frac{\partial}{\partial \theta} \arg h(re^{i\theta}) = \mathrm{Re}\left\{ \frac{re^{i\theta}h'(re^{i\theta})}{h(re^{i\theta})} \right\}.$$

This idea does not readily extend to a higher-dimensional space. Moreover, such an approach is crucially connected with the initial condition $h(0) = 0$. Much later Wald [150] characterized of those functions which are starlike with respect to another center (sometimes these functions are called weakly starlike [65, 66]). Observe that although the classes of starlike, spirallike and convex functions were studied very extensively, little was known about functions that are holomorphic on the unit disk Δ and starlike with respect to a boundary point. In fact, it was only in 1981 that M.S. Robertson [121] introduced two relevant classes of univalent functions and conjectured that they were equal. In 1984 his conjecture was proved by A. Lyzzaik [90]. Finally, in 1990 Silverman and Silvia [134], using a similar method, gave a full description of the class of univalent functions on Δ, the image of which is starshaped with respect to a boundary point. However, the approaches used in their work have a crucially one-dimensional character (because of the Riemann Mapping Theorem and Carathéodory's Theorem on Kernel Convergence). In addition, the conditions given by Robertson and by Silverman and

Silvia, characterizing starlikeness with respect to a boundary point, essentially differ from Wald's and Nevanlinna's conditions of starlikeness with respect to an interior point. Hence, it is difficult to trace the connections between these two closely related geometric objects. Therefore, even in the one-dimensional case the following problem arises: *To find a unified condition of starlikeness (and spiral-likeness) with respect to an interior or a boundary point.*

By 1923 K. Löwner [89] described the problem of continuous deformations of a domain of certain class by using a first order differential equation

$$\frac{\partial u}{\partial t} = \frac{\partial u}{\partial z} \frac{k(t) + u}{k(t) - u} u,$$

$u(0, z) = z$, $z \in \Delta = \{|z| < 1\}$, $0 < t < T$, $|k(t)| = 1$, the right hand of which is time dependent and has a moving polar singularity. This equation is central in the theory of parametric extensions of univalent functions and its application (see, for example [11] and references there). In particular, L. de Branges [30] used it to solve the famous Bieberbach conjecture.

It seems that the idea to use autonomic dynamical systems was first suggested by Robertson [120] in 1961 and developed by Brickman [20] in 1973, who introduced the concept of Φ-like functions as a generalization of starlike and spirallike functions (with respect to the origin) of a single complex variable. Suffridge, Pfaltzgraff [104, 105, 140, 141, 106] and Gurganus [58] (see also [26] and [109]) developed a similar approach in order to characterize starlike, spirallike (with respect to the origin), convex and close-to-convex mappings in higher-dimensional cases. Since 1970 the list of papers on these subjects has became quite long. Nevertheless, it seems that there is no extension of Wald's, as well as of Silverman and Silvia's, results to higher dimensions.

In the last chapter we demonstrate how one can study starlike and spiral-like functions by using the behavior of trajectories defined by related dynamical systems and their characterizations. For example, in the description of starlike functions, this approach, roughly speaking, consists of the following observation: If we consider the flows defined by the evolution equations $du/dt + h(u) = 0$ and $dv/dt + v = 0$, the condition of starlikeness (with respect to a point in the closure of a domain), translates to the fact that h takes integral curves of the first evolution equation into those of the second one. So the problem is to find a suitable condition to describe those holomorphic functions which generate a converging semigroup of holomorphic self-mappings of a domain. Note that such approach can be useful also for higher dimensions (see, for example, [140, 141, 104, 105, 106, 58, 26, 109, 56, 36] and [41]).

My purpose has been to write a tutorial rather than a scientific book. And though many results are previously contained only in journal papers, I have tried to facilitate the understanding of their proofs to students and nonspecialists.

Certain chapters of this book were used in my lectures on Geometrical Theory of Complex Variables at Ort Braude College, Karmiel. More advanced sections were presented in a special course for the College teachers and at the Seminars on Functional Analysis and Nonlinear Analysis at the Technion, Haifa.

It seems that the book will be useful to students and postgraduate students in engineering who may use Complex Dynamic Systems. Furthermore, I hope that it will be of interest to mathematicians specializing in complex analysis and differential equations.

My joint work, as well as numerous talks, with Simeon Reich, Mark Elin and Dov Aharonov convinced me that Generation Theory may serve as a showcase for the Classical Geometrical Function Theory. Simeon Reich has given a series of lectures on Complex Analysis and Hyperbolic Geometry in the Complex Plane. His initiative has stimulated me to design a course on the relationships of various topics related to one-dimensional analysis.

I would like to thank ORT Braude College, Karmiel and the Technion, Haifa for their support throughout the project. I am very grateful to my colleagues Mark Elin, Giora Enden, Yakov Lutsky, Ludmila Shvartsman for their help. Special thanks go to Mark Elin who examined the manuscript in detail and whose proofreading and astute observations led to significant improvements in Chapter V. I am deeply indebted to my wife Tania for typing the final version.

Preliminaries

In this short chapter we compile the series of very basic notions and results, proba-
bly familiar to most readers. As we mentioned in the Preface, only a very modest
preliminary knowledge is required to read the following material. Nevertheless,
certain fundamental topics, such as those related to integral representations, convergence theorems, and fixed point principles, will be used throughout the text,
and therefore should be presented at least an auxiliary material.

0.1 Notations and notions

Throughout the book we shall use the following notation:

$\mathbb{R}$ — the set of all real numbers (*real axis*);

$\mathbb{R}^+ = \{x \in \mathbb{R} : x \geq 0\}$ — the set of all nonnegative real numbers (*the nonnegative real half-axis*);

$\mathbb{C}$ — the set of all complex numbers $z = x + iy$, $x, y \in \mathbb{R}$, $i^2 = -1$ (i.e., *the
complex plane*).

If $z = x + iy \in \mathbb{C}$ then $\bar{z} = x - iy$ is its conjugate complex number, $\operatorname{Re} z :=
x$, $\operatorname{Im} z := y$ and $|z| = \sqrt{x^2 + y^2}$.

As usual, if $z \in \mathbb{C}$ is represented in the form $z = |z|e^{i\theta}$, then $\theta = \arg z$.

$\Pi_+ = \{z \in \mathbb{C} : \operatorname{Re} z \geq 0\}$ — *the right half plane*.

$\Delta_r(a) = \{z \in \mathbb{C} : |z - a| < r\}$ — the open disk of radius $r > 0$ centered at
$a \in \mathbb{C}$.

$\Delta = \Delta_1(0) = \{z \in \mathbb{C} : |z| < 1\}$ — the open unit disk in $\mathbb{C}$.

Definitions of a few technical terms will be given here.

An open connected subset D of $\mathbb{C}$ is called a domain in $\mathbb{C}$. The symbol ∂D will
denote the boundary of D. In particular, $\partial \Delta = \{z \in \mathbb{C} : |z| = 1\}$ is the unit circle.

When z represents any number of a set $D \subset \mathbb{C}$, we call z a complex variable.
The set D is usually a domain in $\mathbb{C}$. If to each value z in D we assign a second
complex variable w, then w said to be a function of the complex variable z on the

set D:

$$w = f(z).$$

The term 'function' signifies a single-valued function unless specified differently.

When $w = f(z)$ and w and z are complex variables some information about f may be conveniently illustrated graphically, however, two separate complex planes for the two variables $z = x + iy$ and $w = u + iv$ are required.

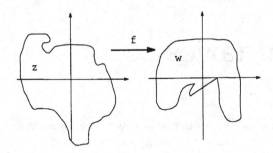

Figure 0.1: The function $w = f(z)$.

The correspondence $w = f(z)$ between points in the two planes is called *a mapping* or *transformation* of points (or sets) given by the function f.

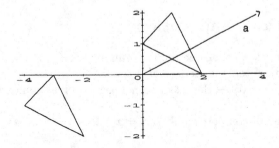

Figure 0.2: The translation $f(z) = z + a$, $a = 4 + 2i$.

We use the following geometrical terms in the form:

translation: $f(z) = z + a$, $a \in \mathbb{C}$, (Figure 0.2)
rotation: $f(z) = e^{i\theta} z$, $\theta \in (0, 2\pi)$, (Figure 0.3)
contraction: $f(z) = kz$ (Figure 0.4).

It is sometimes convenient to consider the mapping as a transformation in a single plane.

Sometimes to conform our terminology with the theory of dynamical systems we will refer to the 'vector field' $f : D \mapsto \mathbb{C}$, as a vector $f(z)$ whose origin is at $z \in D$ (cf., Figure 0.1 and Figure 0.5).

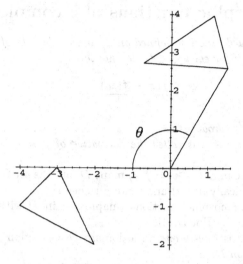

Figure 0.3: The rotation $f(z) = e^{i\pi\theta}z$, $\theta = -2\pi/3$.

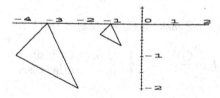

Figure 0.4: The contraction $f(z) = kz$, $k = 1/3$.

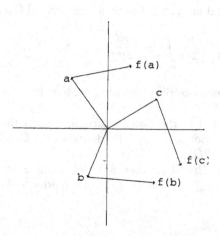

Figure 0.5: The vector field $w = f(z)$.

4

0.2 Holomorphic functions of a complex variable

If f is a complex-valued function defined on an open subset D of $\mathbb{C}$ we say that f is holomorphic in D if for each $z_0 \in D$ the quotient

$$\frac{f(z) - f(z_0)}{z - z_0}$$

has a limit in $\mathbb{C}$ as z approaches z_0.

When this limit exists *it is called the derivative of f at z_0 and is denoted as* $f'(z_0)$.

If f is holomorphic in D and $w = f(z)$ maps D into the set $\Omega \subset \mathbb{C}$ we say that f maps D holomorphically into Ω and write $f : D \mapsto \Omega$.

The set of all holomorphic functions (mappings) in D with values in a set $\Omega \subset \mathbb{C}$ will be denoted by $\mathrm{Hol}(D, \Omega)$.

If such a mapping is *one-to-one* on a domain D a function $f \in \mathrm{Hol}(D, \mathbb{C})$ is said to be *univalent on D.*

In particular, $\mathrm{Hol}(D, \mathbb{C})$ is the vector space of all holomorphic functions in D. We will simply use $\mathrm{Hol}(D)$ to denote the set $\mathrm{Hol}(D, D)$ of holomorphic *self-mappings of D.*

If F and G are in $\mathrm{Hol}(D)$, then $\Phi = F \circ G$ is also in $\mathrm{Hol}(D)$, where $\circ$ denotes the composition operation on this set, i.e., $\Phi(z) = F(G(z))$.

A fundamental property of holomorphic functions is the **Cauchy Integral Formula**:

If $f \in \mathrm{Hol}(D, \mathbb{C})$, then for each closed contour γ in D and each $z \in D \setminus \gamma$

$$\frac{1}{2\pi i} \int_\gamma \frac{f(\zeta)\, d\zeta}{\zeta - z} = \begin{cases} f(z), & \text{if } z \text{ inside } \gamma \\ 0, & \text{otherwise.} \end{cases} \tag{0.2.1}$$

This formula leads to the most basic and important results in the theory of holomorphic functions.

In particular, if $f \in \mathrm{Hol}(D, \mathbb{C})$, then $f'(z)$ also belongs to $\mathrm{Hol}(D, \mathbb{C})$, hence by induction, f is infinitely differentiable.

We adopt the notation $\frac{d^n f}{dz^n}(z)$, for the *n*-th derivative of f at the point $z \in D$. The following integral formula is known as the general Cauchy formula

$$\frac{d^n f}{dz^n}(z) = \frac{n!}{2\pi i} \int_\gamma \frac{f(\zeta) d\zeta}{(\zeta - z)^{n+1}}, \qquad n = 0, 1, 2, \ldots, \tag{0.2.2}$$

where γ here is any closed positively oriented contour in D, such that z is inside γ.

In this formula as well in the Cauchy formula (0.2.1) the contour γ can be replaced by the oriented boundary ∂D of a domain $D \subset \mathbb{C}$, whenever f admits a continuous extension to ∂D.

Furthermore, if $f \in \mathrm{Hol}(D, \mathbb{C})$ then for each point $z_0 \in D$ there is a disk $\Delta_r(z_0) = \{z \in \mathbb{C} : |z - z_0| < r\}$ in D such that f admits *Taylor's power series extension* at z_0:

$$f(z) = \sum_{n=0}^{\infty} a_n(z - z_0)^n, \qquad z \in \Delta_r(z_0),$$

where

$$a_0 = f(z_0), \quad a_n = \frac{1}{n!}\frac{d^n f}{dz^n}(z_0).$$

That is, the infinite series converges to f.

Another very important consequence of the Cauchy integral formula is the so called **maximum modulus principle:**

If $f \in \mathrm{Hol}(D, \mathbb{C})$ is not a constant and $z_0 \in D$ then every neighborhood U of z_0 contains points z such that

$$|f(z)| > |f(z_0)|.$$

In other words, if for $f \in \mathrm{Hol}(D, \mathbb{C})$ the modulus $|f|$ attains a maximum in D, then f is constant.

Further, any real function $u = u(x, y)$ of two real variables x and y that has continuous partial derivatives of the second order and that satisfies Laplace's equation

$$\frac{\partial^2 u}{\partial x^2} + \frac{\partial^2 u}{\partial y^2} = 0$$

throughout a domain D is called a *harmonic* function in D.

For $f \in \mathrm{Hol}(D, \mathbb{C})$ the functions $u = \mathrm{Re}\, f(x + iy)$ and $v = \mathrm{Im}\, f(x + iy)$ are known to be harmonic.

The maximum modulus principle imply **maximum and minimum principles for harmonic functions:**

If $u = u(x, y)$ is a nonconstant harmonic function in D, then it cannot attain neither maximum nor minimum in D.

The function $u = \mathrm{Re}\, f$ plays a crucial role in the study of holomorphic functions on the unit disk Δ.

In particular, each $f \in \mathrm{Hol}(\Delta, \mathbb{C})$ can be expressed in terms of its real part:

$$f(z) = i\,\mathrm{Im}\, f(0) + \frac{1}{2\pi i}\int_\gamma \mathrm{Re}\, f(\zeta)\frac{\zeta + z}{\zeta - z}\frac{d\zeta}{\zeta}, \qquad (0.2.3)$$

where $\gamma = \{\zeta : |\zeta| = r\}$ is a circle of radius $1 > r > |z|$.

This formula is often referred to as the **Cauchy–Schwarz representation formula.**

The holomorphic functions f on Δ whose image lies in the right half-plane Π_+, play a special role in further discussions.

A remarkable result proposed by (0.2.3) and proved by **Herglotz and Riesz** (see, for example [9, 57, 122]) is an **integral representation of positive harmonic functions on Δ:**

if $p \in \mathrm{Hol}(\Delta, \mathbb{C})$ with $p(0) = 1$ and $\mathrm{Re}\, p(z) > 0$, $z \in \Delta$, then

$$p(z) = \int_{|\zeta|=1} \frac{1 + z\bar{\zeta}}{1 - z\bar{\zeta}}\, dm(\zeta), \qquad z \in \Delta,$$

where m is a probability measure on the unit circle.

0.3 Convergence of holomorphic functions

The vector space $\mathrm{Hol}(D, \mathbb{C})$ of holomorphic functions on a domain D is always understood to be endowed with the topology of uniform convergence on compact subsets of D.

Throughout what follows we will use the notation $f_n \to f$, which means that the sequence $\{f_n\} \subset \mathrm{Hol}(D, \mathbb{C})$ converges to f uniformly on every compact subset of D.

Sometimes we will just say that f_n *locally uniformly converges to f*.

The famous **Weierstrass convergence theorem** states that a *local uniform limit of holomorphic functions is holomorphic* (see, for example, [128]).

The family $F \subset \mathrm{Hol}(D, \mathbb{C})$ of holomorphic functions on a domain D is said to be bounded inside D (or locally uniformly bounded) if for each compact subset Ω of D there is a number $M = M(\Omega)$ such that $|f(z)| \leq M$ for all $z \in \Omega$ and for all $f \in F$.

By the celebrated **Montel Theorem** [99, 100] *the family $\mathcal{F}$ is uniformly bounded inside D if and only if it is compact in the topology of local uniform convergence on D, i.e., each sequence $\{f_n\} \subset \mathcal{F}$ contains a subsequence which converges locally uniformly in D.*

Note that a family $\mathcal{F}$ with the property: every sequence in $\mathcal{F}$ contains a subsequence which either converges or diverges to ∞ locally uniformly in D is called a *normal family in D*.

A result of **Montel** [99, 100] asserts:

If $F \subset \mathrm{Hol}(D, \mathbb{C})$ is a normal family and for some $z_0 \in D$ the set $\{f(z_0) : f \in F\}$ is bounded, then F is compact.

In fact, for a compact family the local uniform convergence is equivalent to pointwise convergence. Even more, the **Vitali theorem** [149] states that *if $\{f_n\} \subset \mathrm{Hol}(D, \mathbb{C})$ is bounded inside D and converges for a set of points having a limit point in D, then $\{f_n\}$ converges locally uniformly on D (to some $f \in \mathrm{Hol}(D, \mathbb{C})$).*

Regarding the convergence of univalent functions we recall two theorems which will be useful in the sequel.

The first is a classical **Hurwitz convergence theorem** [68]:

If $\{h_n\}$ is a sequence of univalent function on a domain D which converges to h locally uniformly on D then either h is univalent or constant.

A very powerful tool in Geometric Function Theory is the **Carathéodory Kernel Convergence Theorem** [21]. We will adopt the formulation of this theorem for further purposes.

Let $\{h_n\}$ and h be univalent functions on D with $h_n(0) = h(0)$ and $h'_n(0) > 0$. Then $h_n \to h$ if and only if the following conditions hold:

(i) for every $\omega \in h(\Delta)$ there is a neighborhood Ω such that Ω lies in $h_n(\Delta)$ for n large enough;

(ii) for $\omega \in \partial h(\Delta)$ there is a sequence $\omega_n \in \partial h_n(\Delta)$ such that $\omega_n \to \omega$ as $n \to \infty$.

Note that according to Carathéodory's concept, conditions (i) and (ii) are known for the sequence $\{h_n(\Delta)\}$ to be *convergent with respect to $h(0)$ in the sense of kernel convergence to $h(\Delta)$*.

0.4   Metric spaces and fixed point principles

A metric space is a nonempty set D endowed with a distance function (or metric) $d : D \times D \to \mathbb{R}^+ = [0, \infty)$ which satisfies the following properties:

(i) (symmetry) $d(x, y) = d(y, x)$, for any pair x, y in D;
(ii) (reflexivity) $d(x, y) = 0$ if and only if $x = y$;
(iii) (triangle inequality) $d(x, y) \le d(x, z) + d(y, z)$ for any triple x, y, z in D.

Note that in differential geometry or higher-dimensional complex analysis, such a function d satisfying (i)–(iii) is called 'distance' (or pseudo-distance, if it satisfies (i) and (iii) only) [47], but we prefer to stay with the notation *metric* of the classical functional analysis.

If d is a metric on a domain $D \subset \mathbb{C}$ it is sometimes convenient to call the pair (D, d) a *metric space*. *If each Cauchy sequence (with respect to this metric) in D converges to an element of D then (D, d) is said to be a complete metric space.*

If $a \in D$ the set $B(a, r) = \{x \in D : d(x, a) < r\}$ is called the (open) ball of radius r and (metric) center in a.

The function $d_1(z, w) = |z - w|$, $z, w \in \mathbb{C}$ clearly satisfies properties (i)–(iii) above on each domain D in $\mathbb{C}$ and is called the *euclidean metric* in $\mathbb{C}$. Here balls $B(a, r)$ are actually disks $\Delta_r(a)$ of radius r geometrically centered at $a \in \mathbb{C}$.

In particular, (Δ, d_1) is a metric space. This space, however, is not complete.

Another example of a metric on Δ is the so called *pseudo-hyperbolic metric d_2*, defined as follows

$$d_2(z, w) := \left| \frac{z - w}{1 - z\bar{w}} \right|.$$

It can be shown that (Δ, d_2) is a complete metric space (see Sections 1.1 and 2.1).

The hyperbolic metric on the unit disk

$$d_3(z, w) := \frac{1}{2} \ln \frac{1 + \left| \dfrac{z - w}{1 - z\bar{w}} \right|}{1 - \left| \dfrac{z - w}{1 - z\bar{w}} \right|}$$

is due to Poincaré.

Although the Poincaré metric requires additional computations it has some nice features related to the measure of the lengths in the spirit of Riemann (see Section 2.1).

Constructing a metric that is invariant under a class of mappings is one of the basic tools for a geometrical approach in analysis.

Let (D, d) be a general complete metric space and let F be a self-mapping of D. We will write in this case $F : D \mapsto D$.

If in addition, for each pair x, y in D

$$d(F(x), F(y)) \le d(x, y)$$

we will say that F is d-nonexpansive (or nonexpansive, with respect to d). The term: 'F is a contraction with respect to d' may also be used.

We will see below (Sections 1.1 and 2.1) that *each holomorphic self-mapping of the unit disk Δ is a contraction with respect to metrics d_2 and d_3.*

The central problem in the theory of d-nonexpansive mappings of an underlined metric space (D, d) is finding their fixed points.

A point $a \in D$ is said to be a fixed point of $F : D \mapsto D$ if

$$F(a) = a.$$

A very powerful tool in solving existence problems in many branches of analysis is **The Banach Fixed Point Theorem (or Banach's contraction principle)** which appeared explicitly in his thesis in 1922 (see, for example, [125]).

Let (D, d) be a complete metric space and let $F : D \mapsto D$ be a strict contraction of (D, ρ) in the sense that:

$$d(F(x), F(y)) \leq qd(x, y), \qquad x, y \in D$$

with some $0 < q < 1$.

Then F has a unique fixed point in D. Moreover, for each $x \in D$ the sequence of iterates $\{F^{(n)}(x)\}$ converges to this fixed point.

Here $F^{(1)} = F$, $F^{(2)} = F \circ F$, $F^{(n+1)} = F \circ (F^{(n)})$.

A simple example $Fx = x + 1$, $x \in \mathbb{R}$, shows that the above principle fails in general for nonexpansive mappings.

However, *if (D, d) is a compact metric space, and $F : D \mapsto D$ satisfies the condition*

$$d(F(x), F(y)) < d(x, y)$$

for all x, y in D, then F has a unique fixed point in D and for any $x \in D$ the sequence of iterates $\{F^{(n)}(x)\}$ converges to this point.

Another very popular theorem which guarantees the existence (but not uniqueness) of fixed points in finite-dimensional analysis and algebraic geometry is **Brouwer's Fixed Point Principle** (see, for example [79]):

Each continuous self-mapping of a nonempty bounded closed and convex subset D of $\mathbb{R}^n$ has a fixed point in D.

In the subsequent chapters we will study the iteration theory of holomorphic self-mappings of the unit disk in $\mathbb{C}$ and their influence on the development of geometric function theory and continuous dynamical systems. The Poincaré hyperbolic metric plays a primary role in this study, and therefore deserves some special attention.

Chapter 1

The Wolff–Denjoy theory on the unit disk

There is a long history associated with the problem on iterating holomorphic mappings and their fixed points, the work of H.A. Schwarz (1869), G. Pick (1916), G. Julia (1920), J. Wolff (1926), A. Denjoy (1926) and C. Carathéodory (1929) being among the most important.

1.1 Schwarz–Pick Lemma and automorphisms

Let D be a domain (open connected subset) in the complex plane $\mathbb{C}$, and let $\mathrm{Hol}(D, \mathbb{C})$ be the set of all holomorphic functions (mappings) from D into $\mathbb{C}$.

Throughout what follows we will denote by Δ the unit disk of the complex plane $\mathbb{C}$, and by $\mathrm{Hol}(\Delta)$ the set of holomorphic self-mappings of Δ. This set is a semigroup with respect to the composition operation. The subgroup of $\mathrm{Hol}(\Delta)$ of all holomorphic automorphisms of Δ will be denoted by $\mathrm{Aut}(\Delta)$. In other words, $F \in \mathrm{Hol}(\Delta)$ is an element of $\mathrm{Aut}(\Delta)$ if and only if F^{-1} is well defined on Δ and belongs to $\mathrm{Hol}(\Delta)$.

We begin with the classical Schwarz Lemma which plays a crucial role in geometric function theory.

Proposition 1.1.1 (The Schwarz Lemma) *Let $F \in \mathrm{Hol}(\Delta)$ be such that $F(0) = 0$. Then*

  (i) $|F(z)| \le |z|$ *for all $z \in \Delta$;*
  (ii) $|F'(0)| \le 1$.

  Moreover, if the equality in (i) holds for at least one $z \in \Delta$, $z \ne 0$, or $|F'(0)| = 1$, then $F(z) = e^{i\varphi} z$ for some $\varphi \in [0, 2\pi]$, and thus the equality in (i) holds for all $z \in \Delta$.

Proof. Since $F(0) = 0$ the function F has the following Taylor representation:

$$F(z) = \sum_{k=1}^{\infty} a_k z^k.$$

Therefore the function

$$G(z) = \frac{F(z)}{z} = \sum_{k=1}^{\infty} a_k z^{k-1}$$

is also holomorphic in Δ.

Take any $r \in (0,1)$ and denote $\Delta_r = \{z \in \Delta : |z| < r\}$. By the maximum modulus principle we obtain:

$$|G(z)| = \left|\frac{F(z)}{z}\right| \le \frac{1}{r}, \quad z \in \Delta_r.$$

Letting $r \to 1^-$ we obtain the inequality:

$$|G(z)| \le 1$$

for all $z \in \Delta$, which is equivalent to *(i)*. Note also that $G(0) = F'(0)$, whereby *(ii)* follows too.

If, in addition, $|G(z)| = 1$ for some $z \in \Delta$, then once again by the maximum modulus principle, $G(z) = e^{i\varphi}$ for some $\varphi \in [0, 2\pi]$, hence $F(z) = e^{i\varphi}z$ for all $z \in \Delta$. $\square$

Let us now fix a point $a \in \Delta$ and consider a fractional linear transformation (the so called Möbius transformation) of the unit disk defined by

$$m_a(z) = \frac{z+a}{1+\bar{a}z}. \tag{1.1.1}$$

It is easy to see that the following inequality holds:

$$1 - |m_a(z)|^2 = \frac{(1-|a|^2)(1-|z|^2)}{|1+\bar{a}z|^2} > 0.$$

In turn, this inequality implies that $|m_a(z)| < 1$, whenever $z \in \Delta$ and a is a fixed element of Δ. That is, m_a maps Δ into itself and $m_a \in \mathrm{Hol}(\Delta)$. Since $m_a^{-1} = m_{-a}$ also belongs to $\mathrm{Hol}(\Delta)$, we have that $m_a \in \mathrm{Aut}(\Delta)$.

Now let h be an arbitrary element of $\mathrm{Aut}(\Delta)$. We now claim that h is of the form:

$$h(z) = \lambda m_a(z)$$

for some unimodular number λ. Indeed, if $h(0) = b \in \Delta$, then $m_{-b}(b) = 0$, and

$$g := m_{-b} \circ h$$

satisfies the conditions of the Schwarz Lemma, i.e., $g \in \mathrm{Hol}(\Delta)$ and $g(0) = 0$. Hence $|g(z)| \le |z|$ for all $z \in \Delta$.

On the other hand $g^{-1} = h^{-1} \circ m_b$ also belongs to $\text{Hol}(\Delta)$, and $g^{-1}(0) = 0$. Therefore, $|g^{-1}(w)| \leq |w|$, $w \in \Delta$. Substituting here $w = g(z)$ we obtain $|g(z)| \geq |z|$, hence, in fact, $|g(z)| = |z|$. Consequently $g(z) = e^{i\varphi}z$ for some $\varphi \in [0, 2\pi]$, and

$$h(z) = m_b(g(z)) = \frac{e^{i\varphi}z + b}{1 + e^{i\varphi}z\bar{b}} = e^{i\varphi}\frac{z + e^{-i\varphi}b}{1 + ze^{-i\varphi}\bar{b}}.$$

Setting $a = e^{-i\varphi}b$ and $\lambda = e^{i\varphi}$ we prove our claim. $\square$

Thus we have proved the following assertion.

Proposition 1.1.2 *Each Möbius transformation (1.1.1) is an automorphism of Δ, and each automorphism of Δ is a composition of a Möbius transformation (1.1.1) and a rotation $r_\varphi : r_\varphi(z) = e^{i\varphi}z$, $\varphi \in [0, 2\pi]$.*

Exercise 1. Prove the following rules of compositions ([52], p.63):

$$m_a \circ r_\varphi = r_\varphi \circ m_{r_{-\varphi}(a)} \tag{1.1.2}$$

and

$$m_a \circ m_b = r_\varphi \circ m_{m_b(a)} \tag{1.1.3}$$

with $\varphi = 2\arg(1 + a\bar{b})$.

Exercise 2. Show that the following relations hold: $m_a^{-1} = m_{-a}$ and $\left(-m_{-a}\right)^{-1} = -m_{-a}$.

Exercise 3. Let I denote the identity mapping in $\mathbb{C}$, i.e., $I(z) = z$, $z \in \mathbb{C}$. For a fixed $a \in \Delta$ define a Möbius transformation M_a by the formula $M_a(z) = -m_{-a}(z)$. Show that M_a satisfies the so called involution property: $M_a \circ M_a = I$, i.e., $M_a^{-1} = M_a$.

The invariant form of the Schwarz Lemma is the following assertion.

Proposition 1.1.3 (The Schwarz–Pick Lemma) *Let $F \in \text{Hol}(\Delta)$. Then for each pair z and w in Δ the following inequality holds:*

$$\left| m_{-F(w)}(F(z)) \right| \leq |m_{-w}(z)|$$

or, explicitly,

$$\left| \frac{F(z) - F(w)}{1 - F(z)\overline{F(w)}} \right| \leq \left| \frac{z - w}{1 - z\overline{w}} \right| \quad \text{(the Schwarz–Pick inequality)}.$$

Moreover, if equality holds in this relation for at least one pair $z \neq w$ in Δ then $F \in \text{Aut}(\Delta)$.

Proof. Fix $w \in \Delta$ and consider the mapping $G = m_{-F(w)} \circ F \circ m_w$. It is clear that $G \in \mathrm{Hol}(\Delta)$ and $G(0) = 0$. Then by the Schwarz Lemma we have $|G(z)| \le |z|$, $z \in \Delta$. Replacing z by $m_{-w}(z)$ we obtain the required inequality. If this relation becomes equality for some pair $z \ne w$, then $|G(m_{-w}(z))| = |m_{-w}(z)|$. Again by the Schwarz Lemma $G(z) = e^{i\varphi}z$ for some $\varphi \in [0, 2\pi]$ and for all $z \in \Delta$. Hence G belongs to $\mathrm{Aut}(\Delta)$, and so does $F = m_{F(w)} \circ G \circ m_{-w}$. $\square$

Corollary 1.1.1 *Let $F \in \mathrm{Hol}(\Delta, \mathbb{C})$. Then F maps Δ into its closure $\overline{\Delta}$ if and only if it satisfies the condition:*

$$|F'(z)| \le \frac{1 - |F(z)|^2}{1 - |z|^2}. \tag{1.1.4}$$

Proof. Indeed, if (1.1.4) holds, then $1 - |F(z)|^2 \ge 0$ and $|F(z)| \le 1$.

Conversely. Suppose that $|F(z)| \le 1$. If $|F(z)| = 1$ for some $z \in \Delta$, then F is a constant and (1.1.4) is trivial. Thus we may assume that $|F(z)| < 1$ for all $z \in \Delta$. Now, rewriting the Schwarz–Pick inequality in the form

$$\left| \frac{F(z) - F(w)}{z - w} \right| \le \left| \frac{1 - F(z)\overline{F(w)}}{1 - z\overline{w}} \right|$$

and letting $w \to z$ we obtain (1.1.4). $\square$

Exercise 4. Show that if F is not a constant and (1.1.4) holds as an equality at one $z \in \Delta$, then $F \in \mathrm{Aut}(\Delta)$.

Hint. As in the proof of Proposition 1.1.3, consider the mapping $G = m_{-F(w)} \circ F \circ m_w$. Show that $G'(0) = 1$.

Corollary 1.1.2 (Lindelöf's inequality, [88], p. 11) *If $F \in \mathrm{Hol}(\Delta)$, then it satisfies the following estimate:*

$$|F(z)| \le \frac{|z| + |F(0)|}{1 + |z||F(0)|}, \quad z \in \Delta.$$

Proof. Since

$$1 - |m_{-F(0)}F(z)|^2 = \frac{(1 - |F(0)|^2)(1 - |F(z)|^2)}{|1 + \overline{F(0)}F(z)|^2}$$

we have

$$
\begin{aligned}
|m_{-F(0)}F(z)|^2 &= \left| \frac{F(z) - F(0)}{1 - F(z)\overline{F(0)}} \right|^2 \\
&\ge 1 - \frac{(1 - |F(0)|^2)(1 - |F(z)|^2)}{(1 - |F(0)||F(z)|)^2} \\
&= \frac{(|F(z)| - |F(0)|)^2}{(1 - |F(0)||F(z)|)^2}.
\end{aligned}
$$

On the other hand, it follows by the Schwarz–Pick inequality that

$$\left| \frac{F(z) - F(0)}{1 - F(z)\overline{F(0)}} \right| \leq |z|$$

and

$$\frac{|F(z)| - |F(0)|}{1 - |F(0)||F(z)|} \leq |z|.$$

Solving the last inequality for $|F(z)|$ we obtain our assertion. □

Corollary 1.1.3 (see L.A. Harris [61]) *If $F \in \text{Hol}(\Delta)$ then it satisfies the following estimate*

$$|F(z) - F(0)| \leq |z| \frac{1 - |F(0)|^2}{1 - |F(0)||z|}.$$

Proof. Consider the mapping $G = m_{-F(0)} \circ F$. Solving this equation for F we obtain:

$$F(z) = m_{F(0)}(G(z)) = \frac{G(z) + F(0)}{1 + \overline{F(0)}G(z)}$$

and

$$F(z) - F(0) = \frac{G(z) + F(0) - F(0) - G(z)|F(0)|^2}{1 + \overline{F(0)}G(z)} = G(z) \frac{1 - |F(0)|^2}{1 + \overline{F(0)}G(z)}.$$

But $G(0) = 0$, hence $|G(z)| \leq |z|$, $z \in \Delta$. This implies the required estimate.
□

Remark 1.1.1 Observe also that (1.1.4) implies the estimate:

$$|F'(z)| \leq \frac{1}{1 - |z|^2} \tag{1.1.5}$$

for each $F \in \text{Hol}(\Delta)$ and $z \in \Delta$. Therefore for each $r \in (0,1)$ we obtain the following uniform Lipschitz condition:

$$|F(z) - F(w)| \leq \frac{1}{1 - r^2}|z - w|, \tag{1.1.6}$$

whenever z and w belong to $\Delta_r = \{z \in \Delta : |z| < r\}$. Now if $f \in \text{Hol}(\Delta, \mathbb{C})$ is bounded, i.e., $|f(z)| \leq M$ for all $z \in \Delta$, then

$$F = \frac{1}{M}f \in \text{Hol}(\Delta).$$

Thus we have that *each bounded holomorphic function on Δ is locally uniformly Lipchitzian with respect to Euclidean distance in $\mathbb{C}$.*

Remark 1.1.2 From the point of view of geometric function theory on the unit disk, the Euclidean distance on the disk is inappropriate. The pseudo-hyperbolic distance between points z and w of Δ is the function $d : \Delta \times \Delta \to \mathbb{R}^+ = [0, \infty)$ defined as follows:

$$d(z, w) = |m_{-w}(z)|.$$

It is easy to see that, actually, d is a metric on Δ that induces the usual Euclidean topology.

Thus the Schwarz–Pick Lemma states *the contraction property of a mapping* $F \in \text{Hol}(\Delta)$ *with respect to this metric:*

$$d(F(z), F(w)) \le d(z, w), \quad z, w \in \Delta.$$

Moreover, F is an isometry with respect to d (i.e., $d(F(z), F(w)) = d(z, w)$) if and only if $F \in \text{Aut}(\Delta)$.

Exercise 5. Show that for each $r \in (0, 1)$

$$\lim_{|w| \to 1^-} \inf_{|z| \le r} d(z.w) = 1.$$

Consider now a few geometrical properties of the action of the group $\text{Aut}(\Delta)$ which will be useful in the sequel.

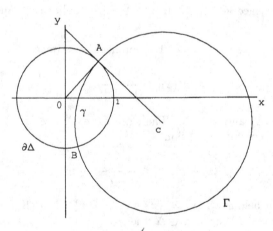

Figure 1.1: An orthogonal circle to $\partial\Delta$ and its image under an automorphism.

Let Γ be a circle orthogonal to $\partial\Delta$, the boundary of Δ, $\Gamma \cap \Delta \ne \emptyset$. If c is the center of Γ, then $|c| \ge 1$, and $\sqrt{|c|^2 - 1}$ is the radius of Γ (see Figure 1.1).

Thus the circle Γ satisfies the equation:

$$\Gamma = \left\{ z \in \mathbb{C} : |z|^2 - 2\,\text{Re}(z\bar{c}) + 1 = 0 \right\}. \tag{1.1.7}$$

Consider now the set $\gamma = \Gamma \cap \Delta$ which is the arc of Γ in Δ with end points A and B (see Figure 1.1). We raise the following question: what is the image of γ under $F \in \text{Aut}(\Delta)$? The next assertion answers this question.

Proposition 1.1.4 *If Γ is an orthogonal circle to $\partial\Delta$, and $\gamma = \Gamma \cap \Delta$, then for each $F \in \text{Aut}(\Delta)$ the image $F(\gamma)$ of γ is either the intersection of Δ with another orthogonal circle to $\partial\Delta$, or it is a diameter of Δ.*

Proof. If F is a rotation of Δ, i.e. $F(z) = e^{i\varphi}z$, $\varphi \in [0, 2\pi]$, $z \in \Delta$, then by (1.1.7) we obtain:

$$F(\gamma) \cap \Delta = \left\{ w \in \mathbb{C} : |w|^2 - 2\,\text{Re}(w\bar{c_1}) + 1 = 0 \right\},$$

where $c_1 = e^{i\varphi}c$ and the proof is completed.

If F is a Möbius transformation m_a, $a \in \Delta$, we have for $w \in m_a(\gamma) \cap \Delta$:

$$\left| m_{-a}(w) - c \right|^2 = R^2 = |c|^2 - 1,$$

or

$$\zeta(|w|^2 + 1) - 2\,\text{Re}(w\bar{\eta}) = 0, \tag{1.1.8}$$

where $\zeta = (1 + 2\,\text{Re}(c\bar{a}) + |a|^2)$ and $\eta = c + c\bar{a}^2 + 2a$. If $\zeta \neq 0$ then (1.1.8) implies:

$$|w|^2 - 2\,\text{Re}(w\bar{c_2}) + 1 = 0,$$

where $c_2 = \eta/\zeta$. Hence $m_a(\gamma) \cap \Delta$ is an intersection of Δ with an orthogonal circle to $\partial\Delta$.

If $\zeta = 1 + 2\,\text{Re}(c\bar{a}) + |a|^2 = 0$, then by (1.1.7) we have that $-a \in \gamma \cap \Delta$. Hence $0 = m_a(-a) \in m_a(\gamma) \cap \Delta$ and once again it follows from (1.1.8) that

$$\text{Re}\,(w\bar{\eta}) = 0.$$

This equation represents a straight line through zero, hence $m_a(\gamma) \cap \Delta$ is a diameter of Δ.

Since each $F \in \text{Aut}(\Delta)$ has the form $e^{i\varphi}m_a$, $\varphi \in [0, 2\pi]$, $z \in \Delta$, (see Proposition 1.1.2.) our proof is completed. $\square$

This assertion states that each automorphism of Δ preserves the class of subsets of Δ which consists of all diameters of Δ and all arcs of the circles which are orthogonal to the boundary of Δ.

We will see below that this result allows us to define the so called geodesic segments on Δ, whose 'length' is preserved under automorphisms.

Another useful image property of automorphisms is the following.

Let the mapping $m_a : \Delta \mapsto \Delta$ be defined as above by formula (1.1.1).

Proposition 1.1.5 *The images of concentric disks $\Delta_r = \{z \in \Delta : |z| < r\}$, $r \in (0,1)$, under the Möbius transformations m_a, $a \in \Delta$, are disks in Δ centered at*

$$c = c(r) = \frac{1 - r^2}{1 - r^2|a|^2} \cdot a \tag{1.1.9}$$

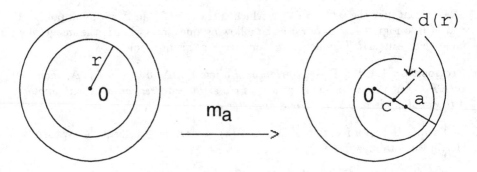

Figure 1.2: A Möbius transformation of Δ_r.

with radius

$$d = d(r) = \frac{1 - |a|^2}{1 - r^2|a|^2} \cdot r. \tag{1.1.10}$$

Proof. Let $\Omega_r(a) := m_a(\Delta_r)$. Then $\Delta_r = m_{-a}(\Omega_r(a))$. In other words, $\Omega_r(a) = \{w \in \Delta : |m_{-a}(w)| < r\}$. So, we need to solve the inequality

$$\left| \frac{w - a}{1 - \bar{a}w} \right| < r.$$

By simple calculations we obtain:

$$\left| w - \frac{(1 - r^2)}{1 - r^2|a|^2} \cdot a \right| < \frac{1 - |a|^2}{1 - r^2|a|^2} \cdot r. \tag{1.1.11}$$

Using notations (1.1.9) and (1.1.10) we obtain our assertion. $\square$

Remark 1.1.3 Thus Proposition 1.1.5 and the Schwarz–Pick Lemma (Proposition 1.1.3) describe the behaviour of holomorphic self-mappings of Δ on the domains $\Omega_r(a)$, $r \in (0,1)$, when a is an interior point of Δ. Namely, the Schwarz–Pick inequality means that for each $F \in \text{Hol}(\Delta)$

$$F(\Omega_r(a)) \subseteq \Omega_r(F(a)), \quad r \in (0,1), \quad a \in \Delta, \tag{1.1.12}$$

i.e., *F maps the disk $\Omega_r(a)$ centered at $c = c(r)$, defined by (1.1.9), into the disk $\Omega_r(F(a))$ centered at*

$$c_1 = \frac{1 - r^2}{1 - r^2|F(a)|^2} \cdot F(a)$$

with the same radius $d = d(r)$, defined by (1.1.10).

However, in fact, the numbers $c = c(r)$ and $d = d(r)$ also depend on the location of the point $a \in \Delta$. In particular, if a approaches the boundary $\partial\Delta$ then for a fixed $r \in (0,1)$ the point c also goes to the boundary while the radius r tends to 0. This is a deficiency in the study of the boundary behavior of holomorphic self-mappings of the unit disk. We will consider this problem in the next section.

Exercise 6. For $a \in \overline{\Delta}$, the closure of Δ, and $K \geq 1 - |a|^2$, define the set

$$D(a, K) = \left\{ z \in \Delta : \frac{|1 - z\bar{a}|^2}{1 - |z|^2} < K \right\}.$$

(a) Show that for an interior point $a \in \Delta$ and $r \in (0,1)$ the sets $\Omega_r(a)$ ($= m_a(\Delta_r)$) and $D(a, K)$ with $K = (1 - |a|^2)(1 - r^2)^{-1}$ coincide.

(b) Show that if $a \in \partial\Delta$ — the boundary of Δ, then for each $K > 0$ the set $D(a, K)$ is a disk in Δ, centered at $a/(1 + K) \in \Delta$ with radius $K/(K+1) < 1$.

Remark 1.1.4 To avoid confusion with the symbol $\Omega_r(a)$ we point out that it differs from the symbol $\Delta_r(a)$, which usually denotes the disk centered at a with radius r.

At the same time the sets $\Omega_r(a)$ are balls with respect to the metric defined by pseudo-hyperbolic distance on Δ (see Remark 1.1.2). Thus an interpretation of the Schwarz–Pick Lemma (or inclusion (1.1.13)) is that *a holomorphic mapping of Δ into itself is a contraction in this metric.*

1.2 Boundary behavior of holomorphic self-mappings

In this section we want to trace dynamics of the Schwarz–Pick inequality when the center of a pseudo-hyperbolic ball is pushed out to $\partial\Delta$.

We will present some classical statements which are based on celebrated results of G. Julia in 1920 [71] and C. Carathéodory's contribution in 1929 [22] (see also [23]).

For a point $\zeta \in \overline{\Delta}$ and $K > 1 - |\zeta|^2$ let us define the sets $D(\zeta, K)$ by the formula

$$D(\zeta, K) = \left\{ z \in \Delta : \frac{|1 - z\bar{\zeta}|^2}{1 - |z|^2} < K \right\}. \tag{1.2.1}$$

It is not difficult to see that for an interior point $\zeta \in \Delta$ the set $D(\zeta, K)$, $K > 1 - |\zeta|^2$, is exactly the pseudo-hyperbolic ball $\Omega_r(\zeta)$ centered at the point ζ with radius $r = \sqrt{1 - \frac{1 - |\zeta|^2}{K}}$ (cf., Exercise 6, Section 1.1).

Note, however, that $D(\zeta, K)$ make sense even if ζ is a boundary point. In this case computations show that for each $K > 0$ the set $D(\zeta, K)$ is geometrically a disk in Δ, centered at $\frac{1}{1+K}\zeta$ with radius $\frac{K}{K+1} < 1$, i.e.,

$$D(\zeta, K) = \left\{ z \in \Delta : \left| z - \frac{1}{1+K}\zeta \right| < \frac{K}{K+1} \right\}. \tag{1.2.2}$$

This disk is internally tangent to the boundary of Δ at the point ζ (Figure 1.3). It is called *a horocycle in* Δ.

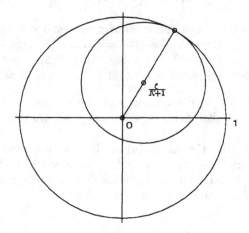

Figure 1.3: A horocycle at the point $\zeta \in \partial\Delta$.

The following assertion establishes the invariance property of horocycles $D(\zeta, K)$ with respect to the family $\mathrm{Hol}(\Delta)$, similarly as was mentioned for domains $\Omega_r(\zeta)$ (see Remark 1.1.3).

Proposition 1.2.1 (Julia's Lemma) *Let* $F \in \mathrm{Hol}(\Delta)$ *and let* $\zeta \in \partial\Delta$ *be a boundary point of* Δ. *Suppose that there exists a sequence* $\{z_n\}_{n=1}^\infty \subset \Delta$ *converging to* ζ *as* n *goes to* ∞, *such that the limits*

$$\alpha = \lim_{n \to \infty} \frac{1 - |F(z_n)|}{1 - |z_n|} \qquad (1.2.3)$$

and

$$\eta = \lim_{n \to \infty} F(z_n) \qquad (1.2.4)$$

exist (finitely).

Then, for each $z \in \Delta$ *the following inequality holds*

$$\frac{|1 - F(z)\bar{\eta}|^2}{1 - |F(z)|^2} \le \alpha \frac{|1 - z\bar{\zeta}|^2}{1 - |z|^2}. \qquad (1.2.5)$$

In other words, for each $K > 0$ *we have the following inclusion:*

$$F\left(D(\zeta, K)\right) \subseteq D(\eta, \alpha K).$$

Proof. It follows that (1.2.4) and (1.2.5) induce $|\eta| = 1$. For $z, w \in \Delta$ we define the function

$$\sigma(z, w) = \frac{(1 - |w|^2)\,(1 - |z|^2)}{|1 - \bar{w}z|^2}. \qquad (1.2.6)$$

It is a simple exercise to show that the Schwarz–Pick inequality (Proposition 1.1.3) can be rewritten in the form

$$\sigma(z, w) \leq \sigma(F(z), F(w)) \tag{1.2.7}$$

for all $F \in \mathrm{Hol}(\Delta)$ and $z, w \in \Delta$. In particular, we have:

$$\sigma(z, z_n) \leq \sigma(F(z), F(z_n))$$

for each $z \in \Delta$ and all $n = 1, 2, \ldots$. Writing the latter inequality explicitly we obtain after simple manipulations:

$$\frac{|1 - \overline{F(z_n)}F(z)|^2}{(1 - |F(z)|^2)} \leq \frac{(1 - |F(z_n)|^2)|1 - \overline{z_n}z|^2}{(1 - |z_n|^2)(1 - |z|^2)}.$$

Letting n go to infinity we obtain (1.2.5). $\square$

Exercise 1. Prove formula (1.2.6) directly.

Exercise 2. For $F \in \mathrm{Hol}(\Delta)$ and $\zeta \in \partial\Delta$, define the value $\alpha(\zeta, F)$ (the Julia number of F at ζ) by the formula:

$$\alpha(\zeta, F) = \liminf_{z \to \zeta} \frac{1 - |F(z)|}{1 - |z|}, \tag{1.2.8}$$

where z tends to ζ unrestrictedly in Δ.

If $\alpha(\zeta, F)$ is finite, it follows by Julia's Lemma, that

$$\frac{|1 - F(z)\bar{\eta}|^2}{1 - |F(z)|^2} \leq \alpha(\zeta, F)\frac{|1 - z\bar{\zeta}|^2}{1 - |z|^2}, \tag{1.2.9}$$

where $\eta = \lim_{n \to \infty} F(z_n)$, $\{z_n\}$ being a sequence along which the lower limit in (1.2.8) is achieved.

Show that if for at least one $z \in \Delta$ we have equality in (1.2.9) then F is an automorphism of Δ.

Remark 1.2.1 Actually inequality (1.2.5) means that *for each $K > 0$, under conditions (1.2.3) and (1.2.4), F maps the horocycle with radius $R = \dfrac{K}{K+1}$ centered at $\dfrac{\zeta}{K+1}$ into the horocycle with radius $r = \dfrac{\alpha K}{\alpha K + 1}$ centered at $\dfrac{\eta}{\alpha K + 1}$* (see formula (1.2.2)). Therefore, if F is a nonconstant holomorphic mapping, then the number α in (1.2.3) must be positive. Also, the same conclusion can be obtain by using the following result of the Schwarz–Pick Lemma.

Exercise 3. Show that for each $F \in \mathrm{Hol}(\Delta)$ and $z \in \Delta$ the following inequality holds:

$$\frac{1 - |F(z)|}{1 - |z|} \geq \frac{1 - |F(0)|}{1 + |F(0)|}.$$

Remark 1.2.2 Observe also, that *if the limit* α *in (1.2.3) exists for a sequence* $z_n \to \zeta \in \partial\Delta$, *then so does limit* $\eta \in \partial\Delta$ *in (1.2.4)*. Indeed, (1.2.5) implies that $|F(z_n)| \to 1$ as $n \to \infty$. Therefore, each subsequence of $\{F(z_n)\}_{n=1}^{\infty}$ has a further subsequence which converges to some unimodular point. Choose two such subsequences and take their limit points, say η_1 and η_2, $|\eta_1| = |\eta_2| = 1$. If $\eta_1 \neq \eta_2$, by decreasing K if necessary, we can find two horocycles $D(\eta_1, \alpha K)$ and $D(\eta_2, \alpha K)$ whose intersection is empty. But from (1.2.5) for each $z \in D(\zeta, K)$ the element $F(z)$ must lie in both of them. Contradiction. Hence $\eta_1 = \eta_2$, and the limit in (1.2.4) exists. Moreover, we will show that, in fact, this limit exists (and equals to η) for each sequence $\{z_n\}$ which converges to ζ along so called nontangential directions. More precisely:

Definition 1.2.1 *For a point* ζ *on the unit circle* $\partial\Delta$ *and* $\kappa > 1$ *a nontangential approach region at* ζ *is the set*

$$\Gamma(\zeta, \kappa) = \{z \in \Delta : |z - \zeta| < \kappa(1 - |z|)\}. \tag{1.2.10}$$

The term 'nontangential' refers to the fact that $\Gamma(\zeta, \kappa)$ lies in a sector S in Δ at point ζ which is the region bounded between two straight lines in Δ that meet at ζ and are symmetric about the radius to ζ, i.e., the boundary curves of $\Gamma(\zeta, \kappa)$ have a corner at ζ, with an intersection angle less than π (see Figure 1.4).

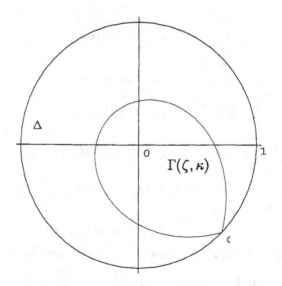

Figure 1.4: A nontangential approach region at a boundary point.

Definition 1.2.2 *We will say that a function* $f \in \mathrm{Hol}(\Delta, \mathbb{C})$ *has a nontangential (or angular) limit* L *at a point* $\zeta \in \partial\Delta$ *if* $f(z) \to L$ *as* $z \to \zeta$, $z \in \Gamma(\zeta, \kappa)$ *for each* $\kappa > 1$. *We will write in this case*

$$L = \angle \lim_{z \to \zeta} f(z).$$

Exercise 4. A Stolz angle at $\zeta \in \partial\Delta$ is the set

$$\widehat{S} = \left\{ z \in \Delta : \left|\arg(1 - \bar{\zeta}z)\right| < \beta, \ |z - \zeta| < r, \ \beta \in (0, \pi/2), \ r \in (0, 2\cos\beta) \right\}$$

(see, Figure 1.5).

Show that $f \in \mathrm{Hol}(\Delta, \mathbb{C})$ has a nontangential limit L at a point ζ if and only if $f(z) \to L$ as $z \to \zeta$, for each Stolz angle $\widehat{S}$ at ζ.

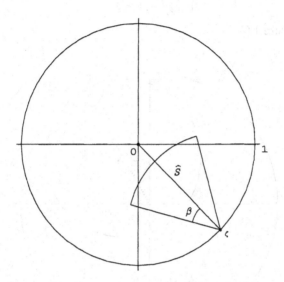

Figure 1.5: A Stolz angle at a boundary point.

A simple consequence of Julia's Lemma is the following.

Corollary 1.2.1 *Let F be a nonconstant holomorphic self-mapping of Δ, and let $\zeta \in \partial\Delta$. Suppose that there is a sequence z_n which converges to ζ, such that*

$$\liminf_{z_n \to \zeta} \frac{1 - |F(z_n)|}{1 - |z_n|} = \alpha < \infty. \qquad (1.2.11)$$

Then:

(i) $\alpha > 0$;

(ii) the nontangential limit

$$\eta := \angle \lim_{z \to \zeta} F(z),$$

which is a point of the boundary $\partial\Delta$ exists;

(iii) for each $K > 0$ the following inclusion holds

$$F(D(\zeta, K)) \subseteq D(\eta, \alpha K).$$

Proof. It remains to show only that condition (1.2.11) implies assertion (ii). Indeed, given $\varepsilon > 0$ choose $K > 0$ such that $D(\eta, \alpha K)$ is contained in the ε-disk centered at η. Further, let S be a sector in Δ with its vertex at ζ. Then one can find $\delta > 0$ such that

$$S_1 = S \cap \{z \in \mathbb{C} : |z - \zeta| < \delta\} \subset D(\zeta, K).$$

Hence by (1.2.5) we have

$$|F(z) - \eta| < \varepsilon$$

for $z \in S_1$ (see figure 1.6) . $\square$

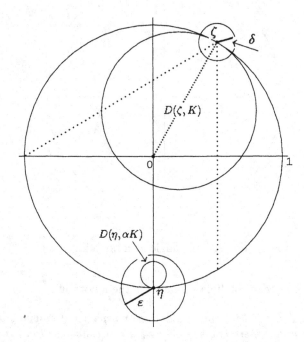

Figure 1.6: Boundary behavior of a self-mapping of Δ.

Finally, we adduce a much stronger assertion, established by Carathéodory (see [22] and [23]).

Proposition 1.2.2 (The Julia–Carathéodory Theorem) *Let $F \in \text{Hol}(\Delta)$ and let $\zeta \in \partial\Delta$. Then the following are equivalent:*

(i) $\liminf\limits_{z \to \zeta} \dfrac{1 - |F(z)|}{1 - |z|} = \alpha < \infty$, *where the limit is taken as z approaches ζ unrestrictedly in Δ;*

(ii) $\angle \lim\limits_{z \to \zeta} \dfrac{F(z) - \eta}{z - \zeta} := \angle F'(\zeta)$ *exists for a point $\eta \in \partial\Delta$;*

(iii) $\angle \lim\limits_{z \to \zeta} F'(z)$ *exists, and* $\angle \lim\limits_{z \to \zeta} F(z) = \eta \in \partial\Delta$.

Moreover,

   (a) $\alpha > 0$ *in (i);*
   (b) the boundary points η in (ii) and (iii) are the same;
   (c) $\angle \lim\limits_{z \to \zeta} F'(z) = \angle F'(\zeta) = \alpha \bar{\zeta} \eta.$

The value $\angle F'(\zeta)$ is called the angular derivative of F at ζ (see also Section 4.6).

There are various proofs of the Julia–Carathéodory Theorem (see for example, the papers [157, 86, 143, 126, 54, 123] and the books [23, 103, 131, 28]. In fact, we will prove below (see Section 4.6) in detail several general assertions which extend Proposition 1.2.2. Nevertheless, in order to demonstrate a direct method and for the completeness we give here a proof which is based on two important classical statements.

Proposition 1.2.3 (Lindelöf's Principle, [88]) *Let $\zeta \in \partial \Delta$ and let function $f \in \mathrm{Hol}(\Delta, \mathbb{C})$ be bounded on each nontangential approach region at ζ. If for some continuous curve $\gamma \in \Delta$ ending at ζ there exists the limit*

$$L = \lim_{z \to \zeta} f(z), \quad z \in \gamma,$$

then the angular limit

$$\angle \lim_{z \to \zeta} f(z) = L$$

also exists.

Proposition 1.2.4 (see [107], p. 79) *Let $\zeta \in \partial \Delta$ and let $f \in \mathrm{Hol}(\Delta, \mathbb{C})$. Suppose that the limit*

$$L = \angle \lim_{z \to \zeta} f(z) := f(\zeta)$$

exists (finitely). Then

$$\angle \lim_{z \to \zeta} \frac{f(z) - f(\zeta)}{z - \zeta}$$

exists if and only if

$$\angle \lim_{z \to \zeta} f'(z)$$

exists and both coincide.

For a proof of this proposition see also Section 4.6.

Proof of Proposition 1.2.2. First we note that the implication $(iii) \Rightarrow (ii)$ is obvious due to Proposition 1.2.4. In turn, $(ii) \Rightarrow (i)$ because of the inequality

$$\frac{1 - |F(r\zeta)|}{1 - r} \leq \frac{|\eta - F(r\zeta)|}{|\zeta - r\zeta|},$$

which holds for all $r \in (0, 1)$.

Now if (i) holds, then by Corollary 1.2.1 there exists

$$\angle \lim_{z \to \zeta} F(z) := \eta \in \partial \Delta. \tag{1.2.12}$$

Thus we need to show that (i) and (1.2.12) imply (ii).

To this end we observe that for each $\kappa > 1$ and $z \in \Gamma(\zeta, \kappa)$ we have by Julia's Lemma

$$\frac{|1 - F(z)\bar{\eta}|^2}{1 - |F(z)|^2} \le \alpha \frac{|1 - z\bar{\zeta}|^2}{1 - |z|^2} = \alpha |z - \zeta| \frac{|z - \zeta|}{1 - |z|^2}$$

$$\le \alpha |z - \zeta| \kappa \frac{1 - |z|}{1 - |z|^2} = \alpha \kappa |z - \zeta|.$$

On the other hand

$$\frac{|F(z) - \eta|}{1 + |F(z)|} = \frac{|F(z) - \eta| \, (1 - |F(z)|)}{1 - |F(z)|^2}$$

$$\le \frac{|1 - F(z)\bar{\eta}|^2}{1 - |F(z)|^2}.$$

This implies that

$$\frac{|F(z) - \eta|}{|z - \zeta|} \le \alpha \kappa \, (1 + |F(z)|) \le 2\alpha\kappa,$$

whenever $z \in \Gamma(\zeta, \kappa)$. In other words, the function $f(z) := \dfrac{F(z) - \eta}{z - \zeta}$ is bounded on each nontangential approach region at ζ. Therefore, to complete the proof, it is sufficient to show (by using Proposition 1.2.3) the equality

$$\lim_{r \to 1^-} \frac{\eta - F(r\zeta)}{\zeta - r\zeta} = \alpha\bar{\zeta}\eta.$$

or, equally,

$$\lim_{r \to 1^-} \frac{\eta - F(r\zeta)}{1 - r} = \alpha\eta. \tag{1.2.13}$$

Indeed, since α is the lower limit in (i) we have

$$\lim_{r \to 1^-} \frac{|\eta - F(r\zeta)|}{1 - r} \ge \lim_{r \to 1^-} \frac{1 - |F(r\zeta)|}{1 - r} \ge \alpha. \tag{1.2.14}$$

On the other hand, setting $z = r\zeta$ in (1.2.5) we obtain

$$\lim_{r \to 1^-} \frac{|\eta - F(r\zeta)|^2}{(1 - r)^2} \le \alpha \lim_{r \to 1^-} \frac{1 - |F(r\zeta)|^2}{1 - r^2}$$

$$= \alpha \lim_{r \to 1^-} \frac{1 - |F(r\zeta)|}{1 + |F(r\zeta)|} \frac{(1 + |F(r\zeta)|)^2}{(1 + r)^2} \frac{1 - r^2}{(1 - r)^2}$$

$$= \alpha \lim_{r \to 1^-} \frac{1 - |F(r\zeta)|}{1 + |F(r\zeta)|} \frac{1 - r^2}{(1 - r)^2}$$

$$= \alpha \lim_{r \to 1^-} \frac{(1 - |F(r\zeta)|)^2}{1 - |F(r\zeta)|^2} \frac{1 - r^2}{(1-r)^2}$$

$$\leq \alpha \lim_{r \to 1^-} \frac{|\eta - F(r\zeta)|^2}{1 - |F(r\zeta)|^2} \frac{1 - r^2}{(1-r)^2}$$

$$\leq \alpha^2 \lim_{r \to 1^-} \frac{|1 - r\zeta\bar{\zeta}|^2}{1 - r^2} \frac{1 - r^2}{(1-r)^2} = \alpha^2. \qquad (1.2.15)$$

Hence, we obtain from (1.2.14) and (1.2.15)

$$\lim_{r \to 1^-} \frac{|\eta - F(r\zeta)|}{1 - r} = \lim_{r \to 1^-} \frac{1 - |F(r\zeta)|}{1 - r} = \alpha \qquad (1.2.16)$$

and

$$\lim_{r \to 1^-} \frac{|\eta - F(r\zeta)|}{1 - |F(r\zeta)|} = \lim_{r \to 1^-} \frac{|1 - \bar{\eta}F(r\zeta)|}{1 - |F(r\zeta)|} = \lim_{r \to 1^-} \frac{|1 - \bar{\eta}F(r\zeta)|}{\mathrm{Re}(1 - \bar{\eta}F(r\zeta))} = 1. \qquad (1.2.17)$$

Thus by (1.2.16) we can write

$$\lim_{r \to 1^-} \frac{\eta - F(r\zeta)}{1 - r} = \alpha\eta e^{i\varphi}, \qquad (1.2.18)$$

where

$$\varphi = \lim_{r \to 1} \arg(1 - \bar{\eta}F(r\zeta)).$$

But (1.2.17) implies that $\varphi = 0$ and we obtain (1.2.13) from (1.2.18). Thus the proof is completed. $\square$

1.3 Fixed points of holomorphic self-mappings

The Schwarz–Pick Lemma implies that if $\zeta \in \Delta$ is an interior fixed point of $F \in \mathrm{Hol}(\Delta)$, i.e.,

$$F(\zeta) = \zeta, \qquad (1.3.1)$$

then F leaves each pseudo-hyperbolic ball $\Omega_r(\zeta)$ centered at ζ invariant.
In other words, for each $r \in (0,1)$,

$$F(\Omega_r(\zeta)) \subseteq \Omega_r(\zeta), \qquad (1.3.2)$$

where

$$\Omega_r(\zeta) = \left\{ z \in \Delta : \left| \frac{z - \zeta}{1 - \bar{\zeta}z} \right| < r \right\} = \left\{ z \in \Delta : \frac{|1 - z\bar{\zeta}|^2}{1 - |z|^2} < K \right\}, \qquad (1.3.3)$$

with $K = (1 - |\zeta|^2)(1 - r^2)^{-1}$ (see Section 1.1, Exercise 6).

In turn, this result shows that a holomorphic self-mapping of Δ which is not the identity has at most one interior fixed point in Δ (see Proposition 1.3.4).

An additional consequence of the Schwarz–Pick Lemma is that if $\zeta \in \Delta$ is a fixed point of Δ, then

$$|F'(\zeta)| \leq 1. \tag{1.3.4}$$

Moreover, the equality in (1.3.2) or (1.3.4) holds if and only if F is an automorphism of Δ.

These facts are helpful in the study of the asymptotic behavior of the discrete time semigroup defined by iterates of a holomorphic self-mapping of Δ (see Proposition 1.3.2).

The situation becomes more complicated if $F \in \mathrm{Hol}(\Delta)$ has no fixed points inside Δ.

If $\zeta \in \partial\Delta$, the boundary of Δ, one can define it as a boundary fixed point of F by the relation

$$\lim_{r \to 1^-} F(r\zeta) = \zeta. \tag{1.3.5}$$

However, simple examples show that holomorphic self-mappings of Δ may have many fixed points on the circle $\partial\Delta$. We begin first with the case of $F \in \mathrm{Aut}(\Delta)$.

1. Fixed points of automorphisms. We already know that if $F \in \mathrm{Aut}(\Delta)$, then it can be presented in the form:

$$F(z) = e^{i\varphi} m_{-a}(z) = e^{i\varphi} \frac{z - a}{1 - \bar{a}z} \tag{1.3.6}$$

for some $a \in \Delta$ and $\varphi \in \mathbb{R}$. Hence, for such F, equation $F(z) = z$ is equivalent to the quadratic equation:

$$\bar{a}z^2 + (e^{i\varphi} - 1)z - e^{i\varphi}a = 0, \tag{1.3.7}$$

or

$$a\bar{z}^2 + (e^{-i\varphi} - 1)\bar{z} - e^{-i\varphi}\bar{a} = 0. \tag{1.3.8}$$

The simplest situation is when $a = 0$. In this case either F is a rotation about the origin $F(z) = e^{i\varphi}z$, $\varphi \in (0, 2\pi)$, or F is the identity $F(z) = z$, $z \in \Delta$. Then, respectively, either F has exactly one fixed point $\zeta = 0$ in Δ or F has infinitely many fixed points in Δ. If F is not the identity and it is not a rotation about zero then $a \neq 0$, and $z = 0$ is not a root of (1.3.8). So we may multiply this equation by $-e^{i\varphi}/\bar{z}^2$, resulting in:

$$\bar{a}\left(\frac{1}{z}\right)^2 + (e^{i\varphi} - 1)\left(\frac{1}{z}\right) - e^{i\varphi}a = 0.$$

Thus $z \neq 0$ is a root of (1.3.7) if and only if $1/\bar{z}$ is a root of (1.3.7).

Consequently, *(1.3.7) has at most one solution inside Δ. If we assume that (1.3.7) has a unimodular solution, then either it is unique or the second solution has also modulus 1.*

Exercise 1. Show that if ζ_1 and ζ_2 are solutions of equation (1.3.7), then for all $z \in \Delta \setminus \{\zeta_2\}$ the following relation holds:

$$\frac{F(z) - F(\zeta_1)}{F(z) - F(\zeta_2)} = \frac{1 - \bar{a}\zeta_2}{1 - \bar{a}\zeta_1} \frac{z - \zeta_1}{z - \zeta_2}, \tag{1.3.9}$$

where $F \in \mathrm{Aut}\,(\Delta)$ is defined by (6).

Thus we see that for $F \in \mathrm{Aut}\,(\Delta)$ the following three situations arise according to the location of its fixed points:

(i) F has exactly one fixed point in Δ;
(ii) F has exactly one fixed point on $\partial\Delta$ and no fixed points in Δ;
(iii) F has two different fixed points on $\partial\Delta$.

The automorphisms of Δ are classified according to these situations, respectively:

In (i) F is said to be elliptic;
In (ii) F is said to be parabolic;
In (iii) F is said to be hyperbolic.

Applying the Schwarz Lemma it is easy to see that an elliptic automorphism F of Δ has the form:

$$F = m_\zeta \circ r_\varphi \circ m_{-\zeta},$$

where $\zeta \in \Delta$ is the solution of (1.3.1) and r_φ is a rotation about zero. Therefore, by Proposition 1.1.5, F is a 'rotation' about ζ (see Figure 1.7).

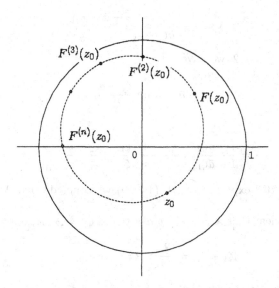

Figure 1.7: Elliptic automorphism.

The dynamical behaviors of parabolic and hyperbolic automorphisms are presented on the Figures 1.8 and 1.9.

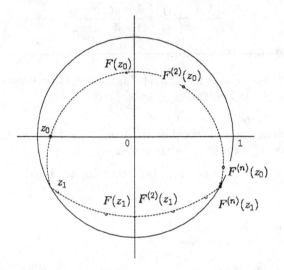

Figure 1.8: Hyperbolic automorphism.

To explain those behaviors we first assume that F, defined by (1.3.6), has two different fixed points, say ζ_1 and ζ_2, $\zeta_1 \neq \zeta_2$, on $\partial\Delta$.

Set
$$\lambda := \frac{1 - \bar{a}\zeta_2}{1 - \bar{a}\zeta_1} = \frac{\overline{\zeta_2 - a}}{\overline{\zeta_1 - a}} \cdot \frac{\zeta_2}{\zeta_1}. \tag{1.3.10}$$

Since $\zeta_j = F(\zeta_j)$, $j = 1, 2$, we have
$$1 - \bar{a}\zeta_j = e^{i\varphi} \frac{\zeta_j - a}{\zeta_j}.$$

Hence
$$\lambda = \frac{1 - \bar{a}\zeta_2}{1 - \bar{a}\zeta_1} = \frac{\zeta_2 - a}{\zeta_1 - a} \cdot \frac{\zeta_1}{\zeta_2} = \frac{\zeta_2 - a}{\zeta_1 - a} \cdot \frac{\overline{\zeta_2}}{\overline{\zeta_1}}.$$

Comparing the latter expression with (1.3.10) we conclude that $\lambda = \bar{\lambda}$ is a real number.

Further, it is clear that $\lambda \neq 1$, since $a \neq 0$ and $\zeta_1 \neq \zeta_2$. Also $\lambda \neq -1$, because otherwise
$$|\zeta_1 + \zeta_2| = \frac{2}{|\bar{a}|} > 2 = |\zeta_1| + |\zeta_2|.$$

In other words, we have shown that $|\lambda| \neq 1$.

Exercise 2. Show that λ, defined by (1.3.10), is positive.

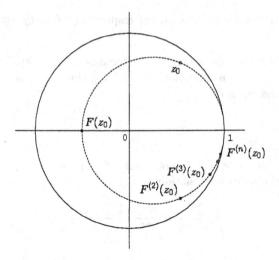

Figure 1.9: Parabolic automorphism.

If we now introduce the fractional linear transformation $L : \mathbb{C} \mapsto \mathbb{C}$ defined by the formula:

$$L(z) = \frac{z - \zeta_1}{z - \zeta_2},$$

then, by using relation (1.3.9) we obtain:

$$L\left(F(z)\right) = \lambda L(z),$$

and consequently:

$$F(z) = L^{-1}\left(\lambda L(z)\right).$$

This implies by induction that

$$F^{(n)}(z) = L^{-1}\left(\lambda^n L(z)\right), \quad n = 0, 1, 2, \ldots,$$

where $F^{(n)}$ are iterates of $F : F^{(0)}(z) = z$, $F^{(n)}(z) = F(F^{(n-1)}(z))$, $n = 1, 2, \ldots$.

Thus, if $|\lambda| < 1$ then for each $z \in \Delta$ the sequence $\left\{F^{(n)}(z)\right\}_{n=0}^{\infty}$ converges to $L^{-1}(0) = \zeta_1$. If $|\lambda| > 1$ then this sequence converges to ζ_2. Since $\left\{F^{(n)}\right\}_{n=0}^{\infty}$ is a normal family on Δ, this convergence is uniform on each compact subset of Δ.

So *if F is a hyperbolic automorphism of Δ, i.e., F has exactly two (distinct) fixed points on the boundary of Δ, then the sequence of iterates $\left\{F^{(n)}\right\}_{n=0}^{\infty}$ converges uniformly on compact subsets of Δ to one of them.*

Moreover, one can estimate such a convergence.

Exercise 3. Prove that for each $z \in \Delta$ the following rate of convergence holds:

$$\left|F^{(n)}(z) - \zeta\right| \leq \frac{4}{1 - |z|}\varepsilon^n,$$

where $\zeta = \zeta_1$ and $\varepsilon = |\lambda|$ if $|\lambda| < 1$, or, respectively, $\zeta = \zeta_2$ and $\varepsilon = |\lambda|^{-1}$ if $|\lambda| > 1$.

Now we consider the case when F is a parabolic automorphism, i.e., F has exactly one fixed point ζ on the boundary of Δ. In this case ζ is a double root of equation (1.3.6) and we have:

$$2\zeta = \frac{1 - e^{i\varphi}}{\bar{a}}.$$

Since $|\zeta| = 1$ and $|a| < 1$ it follows that $e^{i\varphi}$ cannot be -1. Then by direct calculations one can verify the relations:

$$\frac{\zeta}{F(z) - \zeta} = \frac{e^{i\varphi} - 1}{e^{i\varphi} + 1} + \frac{\zeta}{z - \zeta}$$

and

$$\frac{\zeta}{F^{(n)}(z) - \zeta} = n\frac{e^{i\varphi} - 1}{e^{i\varphi} + 1} + \frac{\zeta}{z - \zeta}$$

for all $z \in \Delta$ and $n = 0, 1, 2, \ldots$. The latter equality implies that $F^{(n)}(z)$ converges to $\zeta \in \partial\Delta$ as n tends to infinity.

To summarize our considerations we formulate the following assertion.

Proposition 1.3.1 *If $F \in \mathrm{Aut}(\Delta)$ is not an elliptic automorphism, then the sequence of iterates $\left\{F^{(n)}\right\}_{n=0}^{\infty}$ is convergent. Moreover, if F is not the identity, then the limit of this sequence is a unimodular constant, which is a boundary fixed point of F.*

This result is the first step in the proof of a more general assertion on the asymptotic behavior of holomorphic self-mappings of the unit disk, called the Denjoy–Wolff Theorem.

2. Iterates of holomorphic self-mappings of Δ with an interior fixed point.

Proposition 1.3.2 *Let $F \in \mathrm{Hol}(\Delta)$ have a fixed point $\zeta \in \Delta$. Then:*

(i) for each $r \in (0, 1)$ and $n = 0, 1, 2, \ldots$, the following invariance condition holds:

$$F^{(n)}(\Omega_r(\zeta)) \subseteq \Omega_r(\zeta),$$

where $\Omega_r(\zeta)$ is defined by (1.3.3);

(ii) if F is not the identity then the point $\zeta \in \Delta$ is a unique fixed point of F in Δ.

Moreover, the following are equivalent:

(a) For each $z \in \Delta$, the sequence $\left\{F^{(n)}(z)\right\}_{n=0}^{\infty}$ (the orbit) converges to ζ as n goes to infinity.

(b) The mapping $F : \Delta \mapsto \Delta$ is not an automorphism of Δ.

(c) $|F'(\zeta)| < 1$.

Proof. By the induction method condition (i) is an immediate consequence of inclusion (1.3.2). Also, (1.3.2) implies that $\zeta \in \Delta$ is a unique fixed point of F, if F is not the identity. Indeed, if we assume that F has two different fixed points, say ζ_1 and ζ_2, $\zeta_1 \neq \zeta_2$, then one can choose r_1 and r_2 in $(0,1)$ such that $\zeta_2 \notin \Omega_{r_1}(\zeta_1)$, $\zeta_1 \notin \Omega_{r_2}(\zeta_2)$ and $\overline{\Omega_{r_1}(\zeta_1)} \cap \overline{\Omega_{r_2}(\zeta_2)} = \Omega \neq \emptyset$. It is clear that Ω is a convex closed subset of Δ and $F(\Omega) \subseteq \Omega$. Then it follows by Brouwer's Fixed Point Principle (see Section 0.4) that there exists $\zeta_3 = F(\zeta_3)$ in Ω which is obviously different from ζ_1 and ζ_2. Repeating these arguments we can find a converging sequence $\{\zeta_n\}_{n=1}^{\infty} \subset \Delta$ such that $\zeta_n = F(\zeta_n)$. By the uniqueness property this implies that $F(z) = z$, for all z.

To prove the second part of the assertion we first note that implications (a)$\Rightarrow$(b) and (b)$\Rightarrow$(c) follow directly from the Schwarz–Pick Lemma. Therefore it is enough to prove the implication (c)$\Rightarrow$(a). Since $F'(z)$ is a continuous function on Δ there is a disk $\Delta_r(\zeta) \subset \Delta$ centered at ζ with radius $r > 0$ such that

$$|F'(z)| < 1 \tag{1.3.11}$$

for all $z \in \overline{\Delta_r(\zeta)}$, the closure of $\Delta_r(\zeta)$. In turn, (1.3.11) implies that F satisfies the Lipschitz condition

$$|F(z) - F(w)| \leq q|z - w|, \tag{1.3.12}$$

where $q = \max\left\{|F'(z)|,\ z \in \overline{\Delta_r(\zeta)}\right\}$. In addition, we have from (1.3.12) that F maps $\overline{\Delta_r(\zeta)}$ into itself:

$$|F(z) - \zeta| \leq q|z - \zeta|.$$

So F is a self-mapping of $\overline{\Delta_r(\zeta)}$ which is a strict contraction. It then follows by the Banach Fixed Point Theorem (see Section 0.4) that $\left\{F^{(n)}(z)\right\}_{n=0}^{\infty}$ converges to ζ for all $z \in \overline{\Delta_r(\zeta)}$. Using the Vitali theorem (see Section 0.3) we prove our assertion. $\square$

Combining Proposition 1.3.2. with Brouwer's Fixed Point Principle we obtain the following sufficient condition of existence and uniqueness of an interior fixed point for holomorphic self-mappings of the unit disk.

Corollary 1.3.1 *Suppose that $F \in \mathrm{Hol}(\Delta)$ maps Δ strictly inside, i.e., for some $r \in (0,1)$*

$$|F(z)| \leq r$$

for all $z \in \Delta$. Then F has a unique fixed point $\zeta \in \Delta$, $|\zeta| \leq r$, and for each $z \in \Delta$ the orbit $\left\{F^{(n)}(z)\right\}_{n=0}^{\infty}$ converges to ζ as n goes to infinity.

1.4   Fixed point free holomorphic self-mappings of Δ. The Denjoy–Wolff Theorem.

We will say that F is power convergent if the sequence $S = \left\{F^{(n)}\right\}_{n=1}^{\infty}$ converges uniformly on any subset strictly inside Δ. If the limit of this sequence is a constant $\zeta \in \overline{\Delta}$ then it is called an attractive point of S. Clearly, if ζ is an interior point of Δ then it is a unique fixed point of F.

In this section we intend to study the dynamics of holomorphic self-mappings of Δ with no fixed points inside. A simple case of this situation occurred in the previous section where we saw that an automorphism of Δ with no fixed points has to be either hyperbolic or parabolic, with its fixed point on the boundary of Δ. The content of a remarkable result which was essentially obtained simultaneously by J. Wolff and A. Denjoy is that this fact continues to hold for any holomorphic self-mapping of Δ with no fixed point inside. In other words, each $F \in \mathrm{Hol}(\Delta)$ which has no fixed points in Δ is power convergent to its boundary fixed point in the following sense:

$$\lim_{n \to \infty} F^{(n)}(z) = \zeta \in \partial\Delta$$

and

$$\lim_{r \to 1^-} F(r\zeta) = \zeta.$$

We have already mentioned that a holomorphic self-mapping of Δ may have many fixed points on the boundary $\partial\Delta$ of Δ. So, an additional question is how to recognize which of these is attractive. The key to the answer arrives from Julia's Lemma and the Julia–Carathéodory Theorem where the value of the angular derivative defines such a point (see Proposition 1.4.2 below). Note also that a consequence of the Schwarz–Pick Lemma (Proposition 1.3.2) tell us about the invariance condition in neighborhoods of an interior fixed point of Δ. For mappings with no fixed points a similar result was established by Wolff [157] where pseudo-hyperbolic disks were replaced by horocycles at a certain boundary point of Δ.

Proposition 1.4.1 (Wolff's Lemma) *Let $F \in \mathrm{Hol}(\Delta)$ have no fixed points in Δ. Then there is a unique unimodular point $\zeta \in \partial\Delta$, such that for each $K > 0$ and $n = 0, 1, 2, \ldots$, the horocycle*

$$D(\zeta, K) = \left\{z \in \Delta : \varphi_\zeta(z) := \frac{|1 - z\bar{\zeta}|^2}{1 - |z|^2} < K\right\}$$

internally tangent to $\partial\Delta$ at ζ, is $F^{(n)}$-invariant, i.e.,

$$F^{(n)}\left(D(\zeta, K)\right) \subseteq D(\zeta, K). \tag{1.4.1}$$

Proof. Actually it is sufficient to show that there is a sequence $\{z_n\}_{n=1}^{\infty}$ converging to ζ, which satisfies the conditions of Julia's Lemma, i.e., the limit

(1.2.8):

$$\alpha = \alpha(\zeta, F) := \liminf_{z \to \zeta} \frac{1 - |F(z)|}{1 - |z|}$$

exists. Moreover we will show that $0 < \alpha \leq 1$ and $\angle \lim F(z) = \zeta$.
Then our assertion results as a consequence of Proposition 1.2.2.

Take any arbitrary sequence $\{r_n\}_{n=0}^{\infty} \in (0,1)$ increasing to 1, and consider the mappings $F_n = r_n F$. Then by Corollary 1.3.1 F_n has a fixed point $\zeta_n \in \Delta$. In passing to a subsequence, we may assume that $\{\zeta_n\}_{n=0}^{\infty}$ converges to a point $\zeta \in \overline{\Delta}$. If $\zeta \in \Delta$, then by continuity we have

$$\zeta = \lim_{n \to \infty} \zeta_n = \lim_{n \to \infty} r_n F(\zeta_n) = F(\zeta)$$

contradicting the assumption that F has no fixed point in Δ.
Consequently $|\zeta| = 1$. In addition,

$$\frac{1 - |F(\zeta_n)|}{1 - |\zeta_n|} = \frac{1 - \frac{1}{r_n}|\zeta_n|}{1 - |\zeta_n|} \leq 1$$

for all $n = 1, 2 \ldots$.

Once again, on passing to a subsequence we conclude that the Julia number

$$\alpha(\zeta, F) = \liminf_{z \to \zeta} \frac{1 - |F(z)|}{1 - |z|},$$

(see Section 1.2) exists and is less or equal to 1. It is clear that

$$F(\zeta_n) = \frac{1}{r_n} \zeta_n$$

converges to ζ and by Julia's Lemma we obtain the inequality

$$\frac{|1 - F(z)\bar{\zeta}|^2}{1 - |F(z)|^2} \leq \alpha(\zeta, F) \frac{|1 - z\bar{\zeta}|^2}{1 - |z|^2}, \tag{1.4.2}$$

which implies (1.4.1). $\square$

Exercise 1. Prove the uniqueness of such a point $\zeta \in \partial\Delta$ for which (1.4.2) is satisfied.

Hint: Use the following geometrical property of horocycles as in Figure 1.10 and the invariance property (1.4.2).

We will call such a point $\zeta \in \partial\Delta$ which satisfies the Wolff's Lemma *a sink point of F on $\partial\Delta$.*

Combining Proposition 1.4.1 with Proposition 1.2.2 one obtains a characterization of a fixed point free holomorphic self-mapping of the unit disk Δ which is sometimes called the **Julia–Wolff–Carathéodory Theorem**.

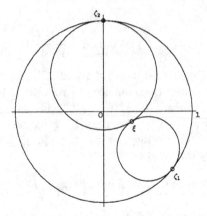

Figure 1.10: Uniqueness of a point ζ.

Proposition 1.4.2 *Let $F \in \text{Hol}(\Delta)$. Then the following are equivalent.*

(i) F has no fixed points in Δ;

(ii) there is a unique unimodular point $\zeta \in \partial\Delta$ such that

$$\alpha := \angle \lim_{z \to \zeta} \frac{F(z) - \zeta}{z - \zeta}$$

exists with $0 < \alpha \leq 1$;

(iii) there is a unique unimodular point $\zeta \in \partial\Delta$ such that

$$\angle \lim_{z \to \zeta} F(z) = \zeta$$

and

$$\angle \lim_{z \to \zeta} F'(z) = \alpha \leq 1;$$

(iv) there is a unique unimodular point $\zeta \in \partial\Delta$ such that

$$\liminf_{z \to \zeta} \frac{1 - |F(z)|}{1 - |z|} = \alpha \leq 1;$$

(v) there is a unique unimodular point $\zeta \in \partial\Delta$ such that

$$\sup_{z \in \Delta} \frac{\varphi_\zeta(F(z))}{\varphi_\zeta(z)} = \alpha \leq 1,$$

where

$$\varphi_\zeta(z) = \frac{|1 - z\bar{\zeta}|^2}{1 - |z|^2} = \frac{|z - \zeta|^2}{1 - |z|^2}, \quad z \in \Delta.$$

Moreover,

(a) the boundary points ζ and the numbers α in (i)–(v) are the same;

(b) the nontangential limits in (ii) and (iii) can be replaced by the radial limits.

Remark 1.4.1 Proposition 1.4.2 (as well as previous Julia's Lemma, the Julia–Carathéodory Theorem and Wolff's Lemma), can be therefore considered as a boundary version of the Schwarz–Pick Lemma; the point ζ is pushed out to $\partial\Delta$ in a suitable manner. The Julia number $\alpha = \alpha(\zeta, F)$ defined by condition (iv) is, in fact, the angular derivative of F at the boundary fixed point ζ, which is also a sink point of F by condition (v). In addition, if $\alpha < 1$, then for each $n = 0, 1, 2, \ldots,$ and all $z \in \Delta$ we have by induction

$$\frac{|F^{(n)}(z) - \zeta|^2}{1 - |F^{(n)}(z)|^2} \leq \alpha^n \frac{|z - \zeta|^2}{1 - |z|^2}. \tag{1.4.3}$$

So $\zeta \in \partial\Delta$ is an attractive boundary fixed point of F, i.e., $F^{(n)}$ — the iterates of F converge to ζ uniformly on each compact subset of Δ with the power rate of convergence (1.4.3) in the sense of the non-Euclidean 'distance' $\varphi_\zeta(z)$ (cf. Corollary 1.3.2).

The question is whether this point is also attractive when $\alpha = 1$. The affirmative answer to this question is given in the next assertion, following Wolff [155, 156, 157] and Denjoy [31].

Proposition 1.4.3 If $F \in \mathrm{Hol}(\Delta)$ has no fixed points in Δ, then there is a unique unimodular point $\zeta \in \partial\Delta$ which is a sink point of F, and the iterates $F^{(n)}$ converge uniformly on each compact subset of Δ to ζ.

Proof. We may assume that $F \notin \mathrm{Aut}(\Delta)$. Since $\{F^{(n)}\}_{n=1}^{\infty}$ is a normal family there is a subsequence $\{F^{(n_j)}\}$ which converges to a holomorphic mapping $G : \Delta \mapsto \overline{\Delta}$. First we show that G must be constant. Indeed, assuming the contrary we have by the maximum modulus principle that $G \in \mathrm{Hol}(\Delta)$. In passing to a subsequence (if necessary), we may assume that the sequences of integers $p_j = n_{j+1} - n_j$ and $q_j = p_j - 1$ tend to infinity, and that the corresponding sequences $\{F^{(p_j)}\}$ and $\{F^{(q_j)}\}$ converge to holomorphic mappings, say h and g, respectively. It follows by the continuity of the composition operation, that

$$h \circ G = \lim_{j \to \infty} \left(F^{(p_j)} \circ F^{(n_j)} \right) = \lim_{j \to \infty} F^{(n_j + 1)} = G.$$

Since G is not a constant, h is also not a constant and it has more than one fixed point in Δ. Hence h must be the identity. At the same time

$$g \circ F = \lim_{j \to \infty} F^{(q_j)} \circ F = \lim_{j \to \infty} F^{(p_j)} = h = I$$

and

$$F \circ g = F \circ \lim_{j \to \infty} F^{(q_j)} = \lim_{j \to \infty} F^{(p_j)} = h = I.$$

Thus $F = g^{-1}$ is an automorphism. Contradiction. Thus $G = \zeta \in \overline{\Delta}$ is a constant.

Now, by Wolff's Lemma (Proposition 1.4.1) ζ must be a sink point of F. Then each convergent subsequence $\{F^{(n_k)}\}$ has the same limit $\zeta \in \partial\Delta$. $\square$

To summarize, we formulate the following result which is sometimes called the Denjoy–Wolff Theorem.

Proposition 1.4.4 *A mapping $F \in \mathrm{Hol}(\Delta)$ is power convergent on Δ if and only if it is not an elliptic automorphism of Δ. Moreover, the limit of the sequence $\left\{ F^{(n)} \right\}_{n=0}^{\infty}$ is a constant $\zeta \in \overline{\Delta}$.*

1.5 Commuting family of holomorphic mappings of the unit disk.

The following result of A.L. Shields [132] is a consequence of the Wolff–Denjoy Theory on the unit disk Δ.

Proposition 1.5.1 *Let Ξ be a commuting family of continuous mappings on $\overline{\Delta}$, such that each $F \in \Xi$ is an element of $\mathrm{Hol}(\Delta)$ Then there is a common fixed point $\zeta \in \overline{\Delta}$ for all $F \in \Xi$.*

Proof. If there is $G \in \Xi$ which has a unique fixed point ζ in $\overline{\Delta}$, then for each $F \in \Xi$ we have:
$$G\left(F(\zeta)\right) = F\left(G(\zeta)\right) = F(\zeta),$$
i.e., $F(\zeta)$ is also a fixed point of G, hence $F(\zeta) = \zeta$.

Suppose now that all mappings in Ξ have at least two different fixed points. This means that for $F \in \Xi$ there is a point $\zeta \in \overline{\Delta}$, such that $\left\{ F^{(n)}(z) \right\}_{n=0}^{\infty}$ converges to ζ for all $z \in \Delta$. But for each $G \in \Xi$:

$$G(\zeta) = \lim_{n \to \infty} G\left(F^{(n)}(z)\right) = \lim_{n \to \infty} F^{(n)}\left(G(z)\right) = \zeta$$

and we have finished. $\square$

Let us consider now a family Ξ_0 of commuting holomorphic self-mappings of Δ which are not necessarily continuous on $\overline{\Delta}$. If at least one element of Ξ_0 has a fixed point in Δ, then this point is unique, hence it is a common fixed point for all mappings in Ξ_0. However, it may be not attractive for all elements of Ξ_0 if Ξ_0 contains an elliptic automorphism of Δ. If all elements of Ξ_0 have no fixed points, then each of them has a sink point on the boundary, which is attractive for a given mapping from Ξ_0. The question is whether there is a common sink point for all elements of Ξ_0? In general, the answer is 'no', as the following example shows.

Example 1. Consider the pair F and G of hyperbolic automorphisms of Δ defined as follows:

$$F(z) = \frac{z-a}{1-\bar{a}z}, \quad G(z) = \frac{z+a}{1+\bar{a}z}, \quad a \in \Delta.$$

It is clear that F and G have the same fixed points ζ_1 and ζ_2 on $\partial\Delta$ and F is commuting with G:

$$F \circ G = G \circ F = I.$$

But if one of these points, say ζ_1, is a sink point of both mappings F and G we will have that for a given point $w \in \Delta$ there is a horocycle $D(\zeta_1, K)$, $K > 0$, such that $w \notin D(\zeta_1, K)$. But there is an integer n such that $F^{(n)}(w) \in D(\zeta_1, K)$ and hence $G^{(n)}\left(F^{(n)}(w)\right) \in D(\zeta_1, K)$. At the same time

$$G^{(n)}(F^{(n)}(w)) = (G \circ F)^{(n)}(w) = w,$$

that is, a contradiction. $\square$

So it does not hold in general that commuting elements of Ξ_0 have the same sink points, but it was shown by D.F. Behan [15] that the only exceptional cases involve pairs of hyperbolic automorphisms of Δ. We will give his result without proof.

Proposition 1.5.2 *If $F \in \Xi_0$ is not a hyperbolic automorphism of Δ and $G \in \Xi_0$ is fixed point free, then F and G have the same sink point on the boundary of Δ.*

Chapter 2

Hyperbolic geometry on the unit disk and fixed points

Another look at the Wolff–Denjoy theory is the using of the so called hyperbolic metric of a domain.

2.1 The Poincaré metric on Δ

Here, we consider some elements of the classical hyperbolic geometry on the unit disk and we will show that the Wolff–Denjoy theory can be extended from $\mathrm{Hol}(\Delta)$ to a wider class of self-mappings of Δ. More detailed representation of the hyperbolic geometry can be found in [55, 47, 52, 32] and [28].

For a class of self-mappings of a domain it is a natural approach to introduce a metric, such that each mapping of this class becomes nonexpansive with respect to such a metric.

For the class of holomorphic mappings there are different ways to define such metrics. For example, one may use the invariant metric defined by the pseudo-hyperbolic distance $d(= d(\cdot, \cdot))$ on Δ (see, section 1.2). However, from the classical Riemann metric theory point of view, it is impossible to measure the 'lengths' of vectors tangent to Δ to obtain d (see details below). It turns out that the only way to eliminate this deficiency is to define the so called Poincaré hyperbolic metric:

$$\rho = \tanh^{-1} d,$$

which inspite of an additional level of computation provides the needed properties in the construction of a non-Euclidean geometry in the spirit of Riemann.

Let z and w be arbitrary points in Δ. Let γ be a circle such that $z, w \in \gamma$, and γ is orthogonal to $\partial \Delta$ (see Figure 2.1).

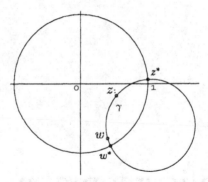

Figure 2.1: The points of anharmonic relation.

Let $z^* \in \gamma \cap \partial\Delta$ and $w^* \in \gamma \cap \partial\Delta$ be such that z^* is on the same side as z and w^* is on the same side as w. The anharmonic relation for these four points is:

$$(z^*, z, w, w^*) = \frac{z - z^*}{z - w^*} \frac{w - w^*}{w - z^*} \tag{2.1.1}$$

Exercise 1. Take a fractional linear transformation g of $\mathbb{C}$,

$$g(z) = (az + b)(cz + d)^{-1}, \; ad - bc \neq 0,$$

and denote

$$u = g(z), \; v = g(w), \; u^* = g(z^*), \; v^* = g(w^*).$$

Show that

$$(z^*, z, w, w^*) = (u^*, u, v, v^*). \tag{2.1.2}$$

Formula (2.1.2) shows that anharmonic relation is invariant under a fractional linear transformation. In particular if we take the Möbius transformation

$$m_{-z}(\xi) = (\xi - z)(1 - \bar{z}\xi)^{-1}$$

we have

$$m_{-z}(z) = 0 \text{ and } m_{-z}(w) = re^{i\varphi} \in \Delta,$$

where $r < 1$, $\varphi \in \mathbb{R}$.

Hence, the line $l = m_{-z}(\gamma)$ is a straight line passing through zero and

$$u^* = m_{-z}(z^*) = -e^{i\varphi}, \; v^* = m_{-z}(w^*) = -e^{i\varphi}.$$

Using (2.1.2) we obtain:

$$(z^*, z, w, w^*) = \frac{e^{i\varphi}}{-e^{i\varphi}} \cdot \frac{re^{i\varphi} - e^{i\varphi}}{re^{i\varphi} + e^{i\varphi}} = \frac{1 - r}{1 + r} = \frac{1 - |m_{-z}(w)|}{1 + |m_{-z}(w)|}. \tag{2.1.3}$$

Proposition 2.1.1 *Let $\rho : \Delta \times \Delta \mapsto \mathbb{R}$ be a function with the following properties:*

(i) for each $F \in \mathrm{Aut}(\Delta)$, and each pair z, w in Δ

$$\rho(F(z), F(w)) = \rho(z, w);$$

(ii) for each pair $s, t \in (0, 1)$, $s < t$

$$\rho(0, t) = \rho(0, s) + \rho(s, t);$$

(iii)
$$\lim_{t \to 0^+} \frac{\rho(0, t)}{t} = 1.$$

Then

$$\rho(z, w) = -\frac{1}{2} \log(z^*, z, w, w^*) = \frac{1}{2} \log \frac{1 + |m_{-z}(w)|}{1 - |m_{-z}(w)|}. \qquad (2.1.4)$$

Proof. First we note that by (i)

$$\rho(z, w) = \rho(m_{-z}(z), \ m_{-z}(w)) = \rho(0, m_{-z}(w)) \qquad (2.1.5)$$

because $m_{-z} \in \mathrm{Aut}(\Delta)$. Also, using the rotation automorphism $r_\varphi : r_\varphi(u) = e^{i\varphi} u$, $u \in \Delta$ with $\varphi = -\arg u$, we have

$$\rho(0, u) = \rho(r_\varphi(0), \ r_\varphi(u)) = \rho(0, |u|). \qquad (2.1.6)$$

Comparing with (2.1.5) we obtain

$$\rho(z, w) = \rho(0, |m_{-z}(w)|). \qquad (2.1.7)$$

So it remains to show that for each $r \in (0, 1)$

$$\rho(0, r) = \tanh^{-1}(r). \qquad (2.1.8)$$

Indeed, for each $\delta > 0$, such that $r + \delta < 1$ we have by (ii) and (i)

$$\begin{aligned}
\rho(0, r + \delta) - \rho(0, r) &= \rho(r, r + \delta) \\
&= \rho(m_{-r}(r), m_{-r}(r, +\delta)) \\
&= \rho\left(0, \frac{\delta}{1 - r\delta - r^2}\right).
\end{aligned}$$

Hence by (iii)

$$\frac{d^+ \rho(0, r)}{dr} = \frac{1}{1 - r^2}.$$

Similarly,

$$\frac{d^- \rho(0, r)}{dr} = \frac{1}{1 - r^2}$$

and we have (2.1.8):

$$\rho(0, r) = \int_0^r \frac{ds}{1 - s^2} = \tanh^{-1}(r)$$

since $\rho(0,0) = 0$. It is also clear that the function $\tanh^{-1}(|m_{-z}(w)|)$ satisfies conditions (i)–(iii). □

Exercise 2. Prove the following property of the real function $\varphi(t) = \tanh^{-1}(t)$:

(a) $\varphi : [0,1) \mapsto [0,\infty)$ is a strictly increasing function which goes to infinity when $t \to 1^-$;

(b) $\dfrac{d\varphi}{dt}\Big|_{t=0} = 1$;

(c) $\varphi\left(\dfrac{t+s}{1+ts}\right) = \varphi(t) + \varphi(s)$.

Proposition 2.1.2 *The function $\rho(\cdot,\cdot)$ defined by (2.1.4) is a metric on Δ.*

Proof. It is clear that $\rho(z,w) \geq 0$ for all $(z,w) \in \Delta \times \Delta$, and since $|m_{-z}(w)| = |m_{-w}(z)|$ we have $\rho(z,w) = \rho(w,z)$ and $\rho(z,w) = 0$ if and only if $z = w$.

Now we need to show the triangle inequality:

$$\rho(z,w) \leq \rho(z,u) + \rho(u,z) \tag{2.1.9}$$

for each triple z, w, u in Δ.

Indeed, let $\gamma : [0,1] \mapsto \Delta$ be a curve joining the points z and w, such that $\gamma'(t)$ is piecewise continuous. We will call such a curve an admissible curve. Consider the function:

$$\delta(z,w) = \inf\{L_\gamma, \ \gamma \text{ is an admissible curve joining } z \text{ and } w \ \},$$

where L_γ is 'length' of γ:

$$L_\gamma = \int_0^1 \frac{|\gamma'(t)|}{1 - |\gamma(t)|^2} dt. \tag{2.1.10}$$

We claim that $\delta(z,w) = \rho(z,w)$. Indeed, it is sufficient to show that $\delta(\cdot,\cdot)$ satisfies the conditions of Proposition 2.1.1. In fact, if $F \in \mathrm{Hol}(\Delta)$ and γ is an admissible curve joining z and w, then $F \circ \gamma$ is an admissible curve joining $F(z)$ and $F(w)$. It follows by Corollary 1.1.1 that

$$
\begin{aligned}
L_{F \circ \gamma} &= \int_0^1 \frac{|F'(\gamma(t))| \cdot |\gamma'(t)|}{1 - |F(\gamma(t))|^2} dt \\
&\leq \int_0^1 \frac{|\gamma'(t)|}{1 - |\gamma(t)|^2} dt = L_\gamma.
\end{aligned}
$$

Hence

$$\delta(F(z), F(w)) \leq \delta(z,w). \tag{2.1.11}$$

If now $F \in \mathrm{Aut}(\Delta)$, then applying (2.1.11) for F^{-1} we obtain the equality:

$$\delta(F(z), F(w)) = \delta(z,w),$$

which proves (i).

To prove (ii) and (iii) it is sufficient to evaluate $\delta(0, s)$, for $0 < s < 1$. If γ joins 0 and s, then so does $\beta = \operatorname{Re} \gamma$. Then we obtain:

$$
\begin{aligned}
L_\gamma &= \int_0^1 \frac{|\gamma'(t)| \, dt}{1 - |\gamma(t)|^2} \\
&\geq \int_0^1 \frac{\beta'(t)}{1 - [\beta(t)]^2} \, dt = \int_0^s \frac{dr}{1 - r^2} = \tanh^{-1}(s),
\end{aligned}
$$

i.e., $\delta(0, s) \geq \tanh^{-1}(s)$. On the other hand, for $\gamma(t) = ts$, $L_\gamma = \tanh^{-1}(s)$ we have:

$$
\delta(0, s) = \tanh^{-1}(s).
$$

Now it follows by properties (b) and (c) of Exercise 2 that function δ satisfies conditions (ii) and (iii) of Proposition 2.1.1.

Thus

$$
\delta(z, w) = \rho(z, w).
$$

Now it can be seen directly that (2.1.9) is a consequence of the definition of $\delta(\cdot, \cdot)$. $\square$

As a matter of fact we establish somewhat more in our proof.

Proposition 2.1.3 *Each mapping $F \in \operatorname{Hol}(\Delta)$ is nonexpansive with respect to the metric ρ, i.e.,*

$$
\rho(F(z), F(w)) \leq \rho(z, w). \tag{2.1.12}
$$

Moreover, if equality in (2.1.12) holds for some pair $z, w \in \Delta$, then $F \in \operatorname{Aut}(\Delta)$.

Actually, the conclusion of this proposition is also a direct consequence of the Schwarz–Pick inequality and (2.1.4).

The metric, whose existence we have just established, is called the Poincaré metric on Δ and will be denoted by ρ.

Proposition 2.1.4 *The pair (Δ, ρ) is a complete unbounded metric space, and ρ defines on Δ the same topology as the original topology on Δ.*

Proof. We list several topological properties of the Poincaré metric which prove our assertion.

(a) $\lim_{n \to \infty} \rho(0, z_n) = \infty$ *if and only if* $|z_n| \to 1^-$.

Indeed, by (2.1.6) we have:

$$
\rho(0, z_n) = \rho(0, |z_n|) = \tanh^{-1} |z_n|
$$

and the assertion follows.

(b) *Each ρ-ball $B(a, R) = \{z \in \Delta : \rho(a, z) < R\}$ is the disk $\Omega_r(a)$ defined by formula (1.2.4), where $r = \tanh R$, i.e.,*

$$
B(a, R) = \{z \in \Delta : |z - sa| < t \cdot \tanh R\}, \tag{2.1.13}
$$

where

$$s = \frac{1 - (\tanh R)^2}{1 - (\tanh R)^2 |a|^2}, \quad t = \frac{1 - |a|^2}{1 - (\tanh R)^2 |a|^2}. \tag{2.1.14}$$

In particular, if $a = 0$, then $B(0, R) = \{z \in \Delta : |z| < \tanh R\}$ is a disk centered at zero.

(c) $B(a, R) = m_a(B(0, R))$, i.e. each ρ-ball is an image of a disk centered at zero under the corresponding Möbius transformation.

Now, it is clear that if $z_n \in \Delta$ converges to $z \in \Delta$ in the sense of ρ-metric, i.e., $\rho(z_n, z) \to 0$, then

$$|z_n - z| \le |z_n - sz| + (1 - s)|z| \to 0$$

because $s \to 1^-$, when $R \to 0$ (see (2.1.14)). Conversely, if $|z_n - z| \to 0$ for $z_n, z \in \Delta$, then there is $r \in (0, 1)$ such that $z_n \in \Delta_r = \{z \in \Delta : |z| \le r\}$.

Hence,

$$\begin{aligned} \rho(z_n, z) &= \tanh^{-1} |m_{-z}(z_n)| \\ &\le \tanh^{-1} \frac{|z_n - z|}{1 - r^2} \to 0. \end{aligned}$$

Finally, note that each ρ-ball is bounded away from the boundary of Δ, and this implies the completeness of the metric space (Δ, ρ). $\square$

2.2   Infinitesimal Poincaré metric and geodesics

Using property (iii) of Proposition 2.1.1 and formula (2.1.7) one can prove a more general assertion than mentioned in part (iii) of Proposition 2.1.1.

Proposition 2.2.1 *Let z be any point of Δ and $u \in \mathbb{C}$ be such that $z + u \in \Delta$. If ρ is the Poincaré metric on Δ, then:*

$$\rho(z, z + u) = \frac{|u|}{1 - |z|^2}(1 + \varepsilon(u)), \tag{2.2.1}$$

where $\varepsilon(u) \to 0$, as $u \to 0$.

Exercise 1. Prove formula (2.2.1).

Formula (2.2.1) shows that the linear differential element $d\rho_z$ in the Poincaré metric is defined by the formula.

$$d\rho_z = \frac{|dz|}{1 - |z|^2}. \tag{2.2.2}$$

Definition 2.2.1 *The form*

$$\alpha(z, u) = \frac{|u|}{1 - |z|^2}, \quad z \in \Delta, \quad u \in \mathbb{C} \tag{2.2.2'}$$

is called the infinitesimal (or differential) Poincaré hyperbolic metric on Δ.

The following properties are an immediate consequence of the definition:

(a) $\alpha(z, u) \geq 0$, $z \in \Delta$, $u \in \mathbb{C}$.
(b) $\alpha(z, tu) = |t| \, \alpha(z, u)$, $t \in \mathbb{C}$.

By Corollary 1.1.1 of the Schwarz–Pick Lemma, each $F \in \text{Hol}(\Delta)$ is a contraction for the infinitesimal Poincaré metric.

Proposition 2.2.2 *If $F \in \text{Hol}(\Delta)$ then $d\rho_{F(z)} \leq d\rho_z$ or equivalently,*

$$\alpha\left(F(z), dF(z)\right) \leq \alpha(z, dz). \tag{2.2.3}$$

If $F \in \text{Aut}(\Delta)$, then the equality in (2.2.3) holds for all $z \in \Delta$. Moreover, if the equality in (2.2.3) holds for at least one $z \in \Delta$, then $F \in \text{Aut}(\Delta)$.

This notion allows us, using Riemann integration, to define the 'length' of any admissible curve in Δ.

Definition 2.2.2 *Let $\gamma : [0,1] \mapsto \Delta$ be an admissible curve in Δ joining two points z and w in Δ. Then the quantity*

$$L_\gamma(= L_\gamma(z, w)) = \int_\gamma d\rho_{\gamma(t)} = \int_0^1 \frac{|\gamma'(t)|}{1 - |\gamma(t)|^2} dt$$

is called the hyperbolic length of γ.

We have already used this notion in Section 1 and have seen that the hyperbolic length is greater than or equal to the hyperbolic distance between its end points, i.e.,

$$\rho(z, w) \leq L_\gamma(z, w). \tag{2.2.4}$$

Definition 2.2.3 *A curve γ joining points z, w in Δ is called a geodesic segment in Δ if its length is equal to the hyperbolic distance between its end points z and w, i.e.,*

$$L_\gamma(z, w) = \rho(z, w). \tag{2.2.5}$$

The following property of geodesics has been discussed, for example, in [47, 52].

Proposition 2.2.3 *For each pair z and w in Δ there is a unique geodesic segment joining z and w and it is either a direct segment, if z and w lie on a diameter of Δ, or a segment of the circle in $\mathbb{C}$ which passes through z and w and is orthogonal to $\partial\Delta$ the boundary of Δ.*

Proof. Indeed, for the points 0 and s, $0 < s < 1$ the curve $\gamma_1(t) = ts$ is a unique curve joining 0 and s such that

$$\rho(0, s) = \int_{\gamma_1} d\rho_{\gamma_1} = \int_0^1 \frac{|\gamma_1'(t)| dt}{1 - |\gamma_1(t)|^2} = \tanh^{-1}(s),$$

i.e., $\gamma_1(t)$ is the geodesic segment joining 0 and s. If z and w are arbitrary points in Δ, then the automorphism $g = r_\varphi \circ m_{-z}$, where m_{-z} is the Möbius transformation and r_φ is the rotation with $\varphi = -\arg m_{-z}(w)$, translates z into 0 and w into $s = |m_{-z}(w)|$.

If we now define $\gamma_1(t)$ as before

$$\gamma_1(t) = t |m_{-z}(w)| \tag{2.2.6}$$

and

$$\gamma(t) = g^{-1} |\gamma_1(t)| \tag{2.2.7}$$

we obtain

$$\gamma(0) = g^{-1}(\gamma_1(0)) = g^{-1}(0) = z$$

and

$$\begin{aligned} \gamma(1) &= g^{-1}(\gamma_1(1)) = g^{-1}(|m_{-z}(w)|) \\ &= m_z(e^{i\varphi}|m_{-z}(w)|) = (m_z \circ m_{-z})(w) = w. \end{aligned}$$

So $\gamma(t)$ is an admissible curve joining z and w.

In addition, by Proposition 2.2.2

$$\begin{aligned} L_\gamma &= \int_0^1 \frac{|\gamma'(t)| dt}{1 - |\gamma(t)|^2} = \int_0^1 \frac{|(g^{-1})'(\gamma_1(t))| \cdot |\gamma_1'(t)|}{1 - n^{-1}(\gamma_1(t))|^2} dt \\ &= \int_0^1 \frac{|\gamma_1'(t)|}{1 - |\gamma_1(t)|^2} dt = L_{\gamma_1}. \end{aligned}$$

But $\rho(z, w) = \rho(0, |m_{-z}(w)|) = L_{\gamma_1}$ and we hence have relation (2.2.5).

The second part of our assertion is a direct consequence of Proposition 1.1.4.

□

2.3 Compatibility of the Poincaré metric with convexity

In this section we will show that the Poincaré metric is compatible with the convex structure of the unit disk Δ.

Proposition 2.3.1 *The Poincaré metric ρ on the unit disk Δ satisfies the following properties:*

(i) If z, w are different points in Δ, then for each $k \in (0,1)$:

$$\rho(kz, kw) < k\rho(z, w).$$

(ii) If z, w and u are three different points in Δ, then for each $k \in (0,1)$

$$\rho((1 - k)z + ku, (1 - k)w + ku) \leq q\rho(z, w),$$

where $q = (1 - k) + k|u| < 1$.

(iii) If z, w, u and v are four points in Δ, then for each $k \in [0,1]$

$$\rho((1 - k)z + ku, (1 - k)w + kv) \leq \max\{\rho(z, w), \rho(u, v)\}.$$

Proof. (i) If γ is the geodesics joining z and w, then $k\gamma$ is an admissible curve joining kz and kw. In addition, calculations show:

$$
\begin{aligned}
\frac{k\gamma'(t)}{1 - k^2|\gamma(t)|^2} &= k\frac{1 - |\gamma(t)|^2}{1 - k^2|\gamma(t)|^2} \cdot \frac{|\gamma'(t)|}{1 - |\gamma(t)|^2} \\
&< k\frac{|\gamma'(t)|}{1 - |\gamma(t)|^2}.
\end{aligned}
$$

Hence,

$$\rho(kz, kw) \leq L_{k\gamma} < kL_\gamma = k\rho(z, w).$$

(ii) Consider a holomorphic mapping F on Δ defined by:

$$F(z) = (1 - k)z + ku,$$

where u is a fixed element of Δ. It follows by the maximum principle that

$$|F(z)| \leq q = (1 - k) + k|u| < 1.$$

Thus

$$G = \frac{1}{q}F \in \text{Hol}(\Delta).$$

Using Proposition 2.1.3 and assertion (i) above we obtain

$$
\begin{aligned}
\rho(qG(z), qG(w)) &= \rho(F(z), F(w)) \\
&:= \rho((1 - k)z + ku, (1 - k)w + ku) \\
&< q\rho(G(z), G(w)) \leq \rho(z, w).
\end{aligned}
$$

(iii) Suppose that

$$\max\{\rho(z, w), \ \rho(u, v)\} = \rho(z, w).$$

Choose F and G in $\text{Aut}(\Delta)$ such that $F(z) = 0$ and $G(u) = 0$, say $F = m_{-z}$ and $G = m_{-u}$.

Denote $|F(w)| = r_1$ and $|G(v)| = r_2$ and take rotations r_{φ_1} and r_{φ_2} with

$$\varphi_1 = -\arg F(w), \quad \varphi_2 = -\arg G(v).$$

Then $F_1 = r_{\varphi_1} \circ F$ and $G_1 = r_{\varphi_2} \circ G$ belong to $\mathrm{Aut}(\Delta)$ and we have

$$F_1(z) = 0, \quad F_1(w) = r_1, \quad G_1(u) = 0, \quad G_1(v) = r_2.$$

Then

$$
\begin{aligned}
\rho(u, v) &= \rho(G_1(u), G_1(v)) = \rho(0, r_2) \\
&\leq \rho(F_1(z), F_1(w)) = \rho(0, r_1),
\end{aligned}
$$

and we obtain $r_2 \leq r_1$.

Denoting the mapping $G^{-1}\left(\dfrac{r_2}{r_1}F\right)$ by Φ we have $\Phi \in \mathrm{Hol}(\Delta)$ and

$$\Phi(z) = G_1^{-1}\left(\frac{r_2}{r_1}F_1(z)\right) = G_1^{-1}(0) = u$$

and

$$\Phi(w) = G_1^{-1}\left(\frac{r_2}{r_1}F_1(w)\right) = G_1^{-1}(r_2) = v.$$

If we consider the mapping $(1-k)I + k\Phi$ which obviously belongs to $\mathrm{Hol}(\Delta)$, again by Proposition 2.1.3 we obtain:

$$
\begin{aligned}
\rho(((1-k)I + k\Phi)(z), ((1-k)I &+ k\Phi)w) \\
&= \rho((1-k)z + k\Phi(z), (1-k)w + k\Phi(w)) \\
&= \rho((1-k)z + ku, (1-k)w + kv) \leq \rho(z, w). \quad \Box
\end{aligned}
$$

Now we continue to study the function $\varphi : [0, 1] \to \mathbb{R}^+$ defined as:

$$\varphi(t) = \rho((1-t)z + tu, (1-t)w + tv) \tag{2.3.1}$$

for given points z, w, u, v in Δ.

Using the equality

$$1 - |m_{-z}(w)|^2 = \sigma(z, w),$$

where

$$\sigma(z, w) = \frac{(1 - |z|^2)(1 - |w|^2)}{|1 - \bar{z}w|^2} \tag{2.3.2}$$

(see section 2.1) we can write $\rho(z, w)$ in the form

$$\rho(z, w) = \tanh^{-1}(1 - \sigma(z, w))^{\frac{1}{2}}. \tag{2.3.3}$$

Therefore, $\rho(z, w) \leq \rho(u, v)$ if and only if $\sigma(z, w) \geq \sigma(u, v)$ and $\sigma(z, w) = 1$ if and only if $z = w$. Also, defining the function $\psi : [0, 1] \mapsto [0, 1]$ by the formula:

$$\psi(t) = \sigma\left((1-t)z + tu, (1-t)w + tv\right) \tag{2.3.4}$$

we have by (2.3.3)

$$\varphi(t) = \tanh^{-1}\sqrt{1 - \psi(t)}. \tag{2.3.5}$$

It is clear that ψ is a continuous function on $[0, 1]$ and $\psi(0) = \sigma(z, w)$. We will show that ψ is right-differentiable at the origin.

Indeed, for $t \neq 0$ we have:

$$\psi(t) = \frac{(1 - |(1-t)z + tu|^2)(1 - |(1-t)w + tv|^2)}{\left|1 - [(1-t)\bar{z} + t\bar{u}][(1-t)w + tv]\right|^2} = \frac{a(t)}{b(t)}, \tag{2.3.6}$$

where

$$
\begin{aligned}
a(t) &= \left[1 - ((1-t)^2|z|^2 + t^2|u|^2 + 2\operatorname{Re}(1-t)t\bar{z}u)\right] \\
&\times \left[1 - ((1-t)^2|w|^2 + t|v|^2 + 2\operatorname{Re}(1-t)t\bar{w}v)\right] \\
&= \left[1 - |z|^2 - (t^2|z|^2 - 2t|z|^2 + t^2|u|^2 + 2\operatorname{Re}(1-t)t\bar{z}u)\right] \\
&\times \left[1 - |w|^2 - (t^2|w|^2 - 2t|w|^2 + t^2|v|^2 + 2\operatorname{Re}(1-t)t\bar{w}v)\right] \\
&= (1 - |z|^2)(1 - |w|^2) \cdot [1 - tc(t)][1 - td(t)], \tag{2.3.7}
\end{aligned}
$$

$$b(t) = \left|1 - [(1-t)\bar{z} + t\bar{u}] \cdot [(1-t)w + tv]\right|^2 \tag{2.3.8}$$

$$c(t) = \frac{(t-2)|z|^2 + t|u|^2 + 2(1-t)\operatorname{Re}\bar{z}u}{(1 - |z|^2)}, \tag{2.3.9}$$

$$d(t) = \frac{(t-2)|w|^2 + t|v|^2 + 2(1-t)\operatorname{Re}\bar{w}v}{(1 - |w|^2)}. \tag{2.3.10}$$

Similarly, we calculate

$$b(t) = |1 - \bar{z}w|^2 \, |1 - te(t)|^2, \tag{2.3.11}$$

where

$$e(t) = \frac{(t-2)\bar{z}w + (1-t)(\bar{z}v + \bar{u}w) + t\bar{u}v}{1 - \bar{z}w}. \tag{2.3.12}$$

So, by (2.3.6)–(2.3.8) we obtain:

$$
\begin{aligned}
\psi(t) &= \frac{(1 - |z|^2)(1 - |w|^2)}{|1 - \bar{z}w|^2} \cdot \frac{(1 - tc(t))(1 - td(t))}{|1 - te(t)|^2} \\
&= \sigma(z, w)\frac{(1 - tc(t))(1 - td(t))}{|1 - te(t)|^2}.
\end{aligned}
$$

This implies

$$\psi'(0^+) = \lim_{t \to 0^+} \frac{1}{t}\left(\psi(t) - \sigma(z, w)\right) = -\sigma(z, w) \cdot \lim_{t \to 0^+} \frac{1}{t}p(t),$$

where

$$p(t) = 1 - \frac{(1 - tc(t))(1 - td(t))}{|1 - te(t)|^2} =$$

$$= \frac{1 - 2t\,\mathrm{Re}\,e(t) + t^2|e(t)|^2 - (1 - tc(t))(1 - td(t))}{|1 - te(t)|^2}$$

$$= \frac{1 - 2t\,\mathrm{Re}\,e(t) + t^2|e(t)|^2 - (1 - t(c(t) + d(t)) + t^2 c(t) \cdot d(t))}{|1 - te(t)|^2}$$

$$= \frac{1 - 2t\,\mathrm{Re}\,e(t) + t^2|e(t)|^2 - 1 + t(c(t) + d(t)) - t^2 c(t) \cdot d(t)}{|1 - te(t)|^2}$$

or

$$\psi'(0^+) = -\sigma(z, w) \cdot \lim_{t \to 0}[-2\,\mathrm{Re}\,e(t) + c(t) + d(t)]$$
$$= -\sigma(z, w) \cdot \lim_{t \to 0^+}[c(t) + d(t) - 2\,\mathrm{Re}\,e(t)].$$

Using (2.3.8), (2.3.9) and (2.3.11) we obtain:

$$\lim_{t \to 0^+} c(t) = \frac{2(\mathrm{Re}\,\bar{z}u - |z|^2)}{1 - |z|^2},$$

$$\lim_{t \to 0^+} d(t) = \frac{2(\mathrm{Re}\,\bar{w}v - |w|^2)}{1 - |w|^2}$$

and

$$\lim_{t \to 0^+} \mathrm{Re}\,e(t) = \mathrm{Re}\,\frac{[(\bar{z}v + \bar{u}w) - 2\bar{z}w]}{1 - \bar{z}w}.$$

Thus we obtain:

$$\psi'(0^+) = -2\sigma(z, w)\left[\frac{\mathrm{Re}\,\bar{z}u - |z|^2}{1 - |z|^2} + \frac{\mathrm{Re}\,\bar{w}v - |w|^2}{1 - |w|^2} - \mathrm{Re}\,\frac{[(\bar{z}v + \bar{u}w) - 2\bar{z}w]}{1 - \bar{z}w}\right].$$
$$(2.3.13)$$

Denoting the expression inside the square brackets of (2.3.13) by μ we see that if $z = w$, then $\mu = 0$. Hence, in this case:

$$\psi'(0^+) = -2\mu\sigma(z, w)$$

also equals to zero.

If $z \neq w$, then by using (2.3.5) and (2.3.13) we obtain:

$$\varphi'(0^+) = \frac{-\psi'(0^+)}{2\sqrt{1 - \psi(0) \cdot \psi(0)}} = \frac{\mu}{|m_{-z}(w)|}.$$
$$(2.3.14)$$

We are now ready to formulate our main result in this section.

Proposition 2.3.2 (cf., [116], Lemma 2.2) *For given four points z, w, u and v in Δ the following statements are equivalent:*

(a) the function $\varphi : [0,1] \mapsto \mathbb{R}^+$, $\varphi(t) = \rho((1-t)z + tu, (1-t)w + tv)$, is not decreasing on $[0,1]$;

(b) $\varphi(0) \leq \varphi(t)$, $\quad t \in [0,1]$;

(c) $\varphi'(0^+) \geq 0$;

(d) $\mathrm{Re} \left[\dfrac{\bar{z}(z-u)}{1 - |z|^2} + \dfrac{\bar{w}(w-v)}{1 - |w|^2} \right] \leq \mathrm{Re} \, \dfrac{\bar{z}(w-v) + w(\bar{z} - \bar{u})}{1 - \bar{z}w}$.

Proof. It is clear that (a) implies (b) which, in turn, implies (c). The equivalence (c) and (d) follows from (2.3.12) and (2.3.13), because $\mu \geq 0$ if and only if (d) holds.

Now, we will show that (c) implies (a). Assume that $\varphi'(0^+) > 0$. We claim that φ is a monotone function on $[0,1]$. Indeed, suppose that φ has a local maximum, which is achieved at the point $t_0 \in (0,1)$. Then there are two points t_1 and $t_2 \in (0,1)$ such that $t_0 = \frac{1}{2}(t_1 + t_2)$ and $\varphi(t_0) > \max\{\varphi(t_1), \varphi(t_2)\}$.

Setting

$$z_1 = (1-t_1)z + t_1 u, \qquad w_1 = (1-t_1)w + t_1 v,$$
$$z_2 = (1-t_2)z + t_2 u, \qquad w_2 = (1-t_2)w + t_2 v,$$

we have $\rho(z_1, w_1) = \varphi(t_1)$ and $\rho(z_2, w_2) = \varphi(t_2)$.

At the same time:

$$\frac{1}{2}z_1 + \frac{1}{2}z_2 = (1-t_0)z + t_0 u,$$
$$\frac{1}{2}w_1 + \frac{1}{2}w_2 = (1-t_0)w + t_0 v.$$

Thus we have by property (iii) of Proposition 2.3.1

$$\begin{aligned} \varphi(t_0) &= \rho\left(\frac{1}{2}z_1 + \frac{1}{2}z_2, \frac{1}{2}w_1 + \frac{1}{2}w_2\right) \\ &\leq \max\{\rho(z_1, w_1), \rho(z_2, w_2)\} \\ &= \max\{\varphi(t_1), \varphi(t_2)\}. \end{aligned}$$

Contradiction. Since $\varphi'(0^+) > 0$ this implies also that φ has no local minimum on $(0,1)$ and therefore φ is monotone. Hence it is not a decreasing function. $\square$

2.4 Fixed points of ρ-nonexpansive mappings on the unit disk

Definition 2.4.1 *A mapping $F : \Delta \mapsto \Delta$, is said to be ρ-nonexpansive if for each pair z, w*

$$\rho(F(z), F(w)) \leq \rho(z, w). \qquad (2.4.1)$$

If for all z, w in Δ

$$\rho(F(z), F(w)) = \rho(z, w), \qquad (2.4.2)$$

then F is called a ρ-isometry of Δ.

It follows by Proposition 2.1.3 that each $F \in \mathrm{Hol}(\Delta)$ is ρ-nonexpansive and $F \in \mathrm{Aut}(\Delta)$ is a ρ-isometry. However, the class of ρ-nonexpansive mappings is essentially wider than $\mathrm{Hol}(\Delta)$.

Indeed, it is not difficult to see that if $F \in \mathrm{Hol}(\Delta)$, then $\overline{F}$ is also ρ-nonexpansive. Therefore, it follows by the Proposition 2.3.1(iii) that each convex combination of F and $\overline{F}$ is also a ρ-nonexpansive. In particular, for $F \in \mathrm{Hol}(\Delta)$, $\mathrm{Re}\, F = \frac{1}{2}(F + \overline{F})$ is a ρ-nonexpansive mapping.

Generally, denoting the class of ρ-nonexpansive mappings by $N_\rho(\Delta)$ we conclude: $N_\rho(\Delta)$ is a closed convex subset of the space of all continuous mappings on Δ, $N_\rho(\Delta)$ is a semigroup with respect to composition, and $\mathrm{Hol}(\Delta)$ is a proper subset of $N_\rho(\Delta)$.

At the same time it can be shown that the set of all ρ-isometries of Δ, usually denoting by $\mathrm{Isom}(\Delta)$ is the joint of $\mathrm{Aut}(\Delta)$ and $\overline{\mathrm{Aut}(\Delta)}$, i.e., $F \in \mathrm{Isom}(\Delta)$ if and only if either $F \in \mathrm{Aut}(\Delta)$ and $\overline{F} \in \mathrm{Aut}(\Delta)$.

We will study here fixed points of ρ-nonexpansive mappings. It will be shown that the Wolff–Denjoy Theory can be extended for the class $N_\rho(\Delta)$. First we note that if F has a fixed point a in Δ, then each ρ-ball $B(a, R) = \{z \in \Delta, \ \rho(z, a) \leq R\}$ is F-invariant. This fact contains Proposition 1.2.1 for holomorphic mappings.

To study more deep properties of fixed points of ρ-nonexpansive mappings we need the following observation. If $F \in N_\rho(\Delta)$, and u is a fixed element of Δ, then the mapping G_t defined by the formula:

$$G_t(z) = (1 - t)u + tF(z), \quad 0 \leq t \leq 1, \qquad (2.4.3)$$

also belongs to $N_\rho(\Delta)$. Moreover, it follows by Proposition 2.3.1(ii) that all $z, w \in \Delta$, $z \neq w$, and $t \in [0, 1)$

$$\rho(G_t(z), \ G_t(w)) < q\rho(z, w), \qquad (2.4.4)$$

where $q = (1 - t)|u| + t < 1$, i.e., G is a strict contraction of (Δ, ρ). Therefore by the Banach Fixed Point Theorem G_t has a unique fixed point $z_t = z_t(u)$ in Δ, and

$$z_t = \lim_{n \to \infty} G_t^{(n)} \qquad (2.4.5)$$

uniformly on all compact subsets of Δ. In addition, it is clear that if for some sequence $\{t_n\}$, $0 < t_n < 1$ there is the limit

$$a = \lim_{t_n \to 1^-} z_{t_n}, \tag{2.4.6}$$

in Δ, then a must be a fixed point of F. Indeed, since $a \in \Delta$ and F is continuous we have:

$$
\begin{aligned}
F(a) - a &= \lim_{t_n \to 1^-} [F(z_{t_n}) - z_{t_n}] \\
&= \lim_{t_n \to 1^-} [F(z_{t_n}) - (1 - t_n)u - t_n(F(z_{t_n}))] \\
&= \lim_{t_n \to 1^-} (1 - t_n)[F(z_{t_n}) - u] = 0.
\end{aligned}
$$

This implies $F(a) = a$.

Definition 2.4.2 ([52]) *For a fixed element $u \in \Delta$ the curve $z_t = z_t(u) : [0, 1) \mapsto \Delta$ defined by formulas (2.4.3) and (2.4.5) is called an approximating curve for F.*

From a different perspective, fixing $t \in [0, 1)$ and varying $u \in \Delta$ we can consider z_t as a mapping from Δ into itself which will be denoted by $J_t = J_t(u)$, that is $J_t : \Delta \mapsto \Delta$ is a mapping implicitly defined by the equation:

$$J_t(u) = (1 - t)u + tF(J_t(u)) \tag{2.4.7}$$

or, in operator form,

$$J_t = (1 - t)I + tF \circ J_t. \tag{2.4.7'}$$

Proposition 2.4.1 *If $F \in N_\rho(\Delta)$, then for each $t \in [0, 1)$ the mapping J_t is also in $N_\rho(\Delta)$ and for $t \in (0, 1)$, the mapping J_t has the same fixed points as F.*

Proof. Let u and v be different points in Δ. For iterates $G_t^{(n)}$, $G_t^{(0)} = I$ we have:

$$
\begin{aligned}
\rho(G_t^{(n+1)}(u), G_t^{(n+1)}(v)) &= \rho((1 - t)u + tF(G_t^{(n)}(u)), (1 - t)v + tF(G_t^{(n)}(v))) \\
&\leq \max\{\rho(u, v), \rho(G_t^{(n)}(u), G_t^{(n)}(v))\} = \rho(u, v)
\end{aligned}
$$

by induction. Hence, from (2.4.5) and (2.4.7) we have:

$$\rho(J_t(u), \quad J_t(v)) \leq \rho(u, v), \tag{2.4.8}$$

i.e., $J_t \in N$ for all $t \in [0, 1)$. Now it is clear that if $a \in \text{Fix}\, J_t$, i.e. a is a fixed point of J_t, then by (2.4.7) we obtain:

$$a = (1 - t)a + tF(a), \quad t \in (0, 1).$$

Assume now that F has a b, which is not a fixed point of J_t for some $t \in (0, 1)$. Then

$$J_t(b) = (1 - t)b + tF(J_t(b)) \neq b.$$

At the same time b is a solution of the equation:

$$z_t = (1 - t)b + tF(z_t).$$

Since the latter equation has a unique solution for each $0 < t < 1$, it follows that $J_t(b)$ must be equal to b. $\square$

Remark 2.4.1 Since the set $\mathrm{Hol}(\Delta)$ is closed with respect to the topology of uniform convergence on each compact subset on Δ, we have by the constructions of Proposition 2.4.1, that *if F belongs to $\mathrm{Hol}(\Delta)$ then so does J_t for each $t \geq 0$.*

We can also obtain the following consequence of this proposition.

Proposition 2.4.2 *If F is a fixed point free ρ-nonexpansive mapping of Δ, then for each $u \in \Delta$ we have:*

$$|J_t(u)| \to 1, \quad as \quad t \to 1^-. \tag{2.4.9}$$

Conversely, if (2.4.9) holds for at least one $u \in \Delta$, then F is fixed point free. In other words, $F \in N$ has a fixed point in Δ if and only if there are $u \in \Delta$ and a subsequence $t_n \to 1^-$ such that the approximating sequence $z_{t_n} = J_{t_n}(u)$ is bounded away from the boundary of Δ.

Proof. Indeed if for some $u \in \Delta$ the sequence $\{J_{t_n}(u)\}$ lies strictly inside Δ, then there is a subsequence which converges to a point $a \in \Delta$. This point must be a fixed point of F, as we saw above.

Now, if F has a fixed point $a \in \Delta$ then $J_t(a) = a$ for all $t \in (0, 1]$ and we obtain by (2.4.8)

$$\rho(J_t(u), \quad J_t(a)) \leq \rho(u, a) < \infty, \quad u \in \Delta.$$

Thus $J_t(u)$ is strictly inside Δ for all $t \in (0, 1)$ contradicting (2.4.9). $\square$

Actually, we can show somewhat more: If F is fixed point free, then there is a unique point $e \in \partial\Delta$, such that for each $u \in \Delta$, the net $\{J_t(u)\}$ converges to e, as t tends to 1^-.

The following assertion contains an extension of Wolff's Lemma.

Proposition 2.4.3 ([52], p. 131) *Let F be a ρ-nonexpansive mapping of the unit disk Δ, and suppose that F has no fixed point in Δ. Then there is a unique point $e \in \partial\Delta$ such that:*
(a) each horocycle

$$D(e, K) = \left\{ z \in \Delta : \frac{|1 - z\bar{e}|^2}{1 - |z|^2} < K, \quad K > 0 \right\}$$

internally tangent to $\partial\Delta$ at e is F-invariant, i.e., $F(D(e, K)) \subseteq D(e, K)$.
(b) For each $t \in (0, 1)$

$$J_t(D(e, K)) \subseteq D(e, K).$$

*(c) For each $u \in \Delta$ the approximating curve $z_t(u) = J_t(u)$ converges to e, as t
tends to 1^-, i.e.,*

$$\lim_{t \to 1^-} z_t(u) = e.$$

Proof. We already know that for each $u \in \Delta$:

$$\lim_{t \to 1^-} |J_t(u)| = 1. \tag{2.4.10}$$

Take firstly $u = 0$ and consider a net $z_t = J_t(0)$. Since $\{z_t\}_{t \in (0,1)}$ is precompact
there is a sequence $z_n = J_{t_n}(0)$ which converges to a point $e \in \overline{\Delta}$. It follows by
(2.4.10) that $|e| = 1$. Since F is ρ-nonexpansive we have:

$$\sigma(F(z_n), F(z)) \geq \sigma(z_n, z) \tag{2.4.11}$$

for each $z \in \Delta$. Recall also that

$$F(z_n) = \frac{1}{t_n} z_n, \tag{2.4.12}$$

and using (2.4.11) and (2.4.12) we obtain:

$$\begin{aligned}
\frac{|1 - \overline{F(z_n)}F(z)|^2}{1 - |F(z)|^2} &\leq \frac{1 - |F(z_n)|^2}{1 - |z_n|^2} \cdot \frac{|1 - \bar{z}_n z|^2}{1 - |z|^2} \\
&= \frac{1 - \frac{1}{t_n}|z_n|^2}{1 - |z_n|^2} \cdot \frac{|1 - \bar{z}_n z|^2}{1 - |z|^2} \\
&< \frac{|1 - \bar{z}_n z|^2}{1 - |z|^2}.
\end{aligned} \tag{2.4.13}$$

In addition (2.4.12) implies that:

$$\lim_{n \to \infty} F(z_n) = e,$$

and letting $n \to \infty$ in (2.4.13) we obtain:

$$\frac{|1 - \bar{e}F(z)|^2}{1 - |F(z)|^2} \leq \frac{|1 - \bar{e}z|^2}{1 - |z|^2}.$$

This means that

$$F(D(e, K)) \subseteq D(e, K)$$

for all $K > 0$. So e is indeed a sink point of F. Its uniqueness can be proved
exactly as in Proposition 1.4.2.

Now, since $D(e, K)$ is convex it follows that for each $t \in [0, 1)$ and $u \in D(e, K)$
the mapping G_t, defined by the formula

$$G_t(z) = (1 - t)u + tF(z)$$

maps $D(e, K)$ into itself, and so does $J_t = \lim_{n \to \infty} G_t^{(n)}$.

Thus for each $t \in (0,1)$ the horocycle $D(e, K)$ is J_t-invariant. Hence for a given $z \in \Delta$ one can find $K > 0$ such that $z \in D(e, K)$ and

$$\frac{|1 - \bar{e}J_t(z)|^2}{1 - |J_t(z)|^2} < K.$$

But together with (2.4.10) this implies that $J_t(z)$ converges to e as t tends to 1^-. Finally, if F is contractive then it follows by the general theory of compact mappings on metric spaces (see, for example, [84]), that $F^{(n)}$ converges to a point $a \in \overline{\Delta}$. Arguments similar to that above show that a must be equal to $e \in \partial \Delta$, and we have completed the proof. $\square$

Remark 2.4.2 Note that the situation is simple when $F \in \text{Isom}(\Delta)$ and it is fixed point free. In this case either F or $\overline{F}$ is an automorphism which is power convergent to a sink point of F on the boundary of Δ. The case of $F^{(n)}$ iterations when F is a fixed point free ρ-nonexpansive mapping neither holomorphic nor antiholomorphic, seems to be rather complicated in the framework of the pure one-dimensional metric theory.

The full answer to this question was given by R. Sine [135] in 1989: *if F is a fixed point free ρ-nonexpansive mapping, then it is power convergent.*

The result for contractive F (i.e., $\rho(F(z), F(w)) < \rho(z, w)$) has established earlier by K. Goebel and S. Reich (see [52], p.138).

Now we turn to the situation when $F \in N_\rho(\Delta)$ has a fixed point in Δ.

Generally, such a situation differs form the holomorphic case. Namely, we mean that there are ρ-nonexpansive mappings which have more than one interior fixed point. The following simple example $F(z) = \frac{1}{2}(z + \bar{z})$ shows that in fact F may have infinite number of fixed points. Actually, in this example F is a ρ-nonexpansive retraction ($F \circ F = F$) of the unit disk onto the open interval $(-1, 1)$. In general, the iterates of a ρ-nonexpansive mapping with interior fixed points may not be convergent. A rather complete discussion on their behavior can also be found in [135].

Thus the following question arises: does the approximating curve $J_t(u)$ for a fixed $u \in \Delta$ converge to a fixed point of F, when t tends to 1^- ?

The following assertion states that the answer to the above question is affirmative.

Proposition 2.4.4 ([52], Theorem 28.3, p. 134) *Let F be a ρ-nonexpansive mapping on Δ with the nonempty fixed point set. Then there exists the limit*

$$\varphi = \lim_{t \to 1^-} J_t$$

which is a retraction onto this set.

This assertion together with assertions (b) and (c) of Proposition 2.4.3 can be considered as an implicit continuous analog of the Wolff–Denjoy Theory for ρ-nonexpansive mappings.

Since J_t for each $t \in (0,1)$ has the same fixed point set as F, one may ask what happens with the discrete iterations $J_t^{(n)}$ for a fixed $t \in (0,1)$. The following assertion is a special case of Theorems 30.5 and 30.8 in [52]. For the holomorphic case, see, also [51].

Proposition 2.4.5 *Let F be a ρ-nonexpansive mapping on the unit disk Δ. Then for each $t \in (0,1)$ the mapping J_t defined by (2.4.3) and (2.4.5) is power convergent. In particular, if F has no fixed point in Δ, then the iterates $J_t^{(n)}(z)$, $n = 1,2\ldots$, converge to the sink point $e \in \partial\Delta$ of F as $n \to \infty$ for each $t \in (0,1)$ and each $z \in \Delta$.*

This assertion is quite simple if F is holomorphic. Indeed, it follows by algorithm (2.4.5) that for each $t \in (0,1)$ the mapping J_t belongs to $\mathrm{Hol}(\Delta)$. Thus to prove our assertion we need only to show that for every $F \in \mathrm{Hol}(\Delta)$ and $t \in (0,1)$, the mapping J_t cannot be an elliptic automorphism. In fact, if $a \in \Delta$ is the fixed point of F then it follows by the chain rule that

$$(J_t)'(a) = (1-t) + tF'(a)(J_t)'(a),$$

i.e.,

$$(J_t)'(a) = \frac{1-t}{1-tF'(a)}.$$

If $F'(a) = 1$ then F, hence J_t, is the identity for each $t \in (0,1)$ and we have completed the proof.

If $F'(a) \neq 1$, then it follows by the Schwarz–Pick Lemma (Proposition 1.1.3(ii)) that $|F'(a)| \leq 1$ and we have:

$$|(J_t'(a)| < 1.$$

This means that J_t is not an automorphism of Δ. $\square$

For a complete exposition of the Wolff–Denjoy Theory under the hyperbolic geometry approach it would be appropriate to mention the results due to P. R. Mercer [95, 97, 98], R. Sine [135] and P. Yang [160] for the one-dimensional case, M. Abate [1]–[5], G.-N.Chen [24], Y. Kubota [80] and P. R. Mercer [96] for the finite-dimensional case, T. Kuczumov and A. Stachura [82], [83], P. Mazet and J. P. Vigué [92], [93], S. Reich [111]–[113], S. Reich and I. Shafrir [114], I. Shafrir [129, 130] for the Hilbert ball and its product, and P. Mellon [94] and K. Wlodarczyk [152]–[153] for $\mathbf{J}^*$-algebras (see also a survey of S. Reich and D. Shoikhet [118]).

P. Yang in [160] suggested a hyperbolic metric characterization of the horocycles $D(e,K)$ in Δ as follows:

$$D(e,K) = \left\{ z \in \Delta : \lim_{z \to e} [\rho(z,w) - \rho(0,w)] < \frac{1}{2}\log K \right\}. \qquad (2.4.14)$$

Even for holomorphic case this characterization is a very useful tool in the study of the boundary behavior of self-mappings of Δ.

By using (2.4.14) and a strengthened Schwarz–Pick inequality, P. R. Mercer [98] has recently obtained a sharper description of Julia's Lemma which makes the Denjoy–Wolff Theorem more transparent. We state his result here.

Proposition 2.4.6 *Let $F \in \mathrm{Hol}(\Delta)$ be not a constant. Suppose that there exists a unimodular point $e \in \partial\Delta$ such that*

$$\liminf_{z \to e} \frac{1 - |F(z)|}{1 - |z|} = \alpha < \infty.$$

Then there is a point η on the boundary $\partial\Delta$, such that for any $z \in \partial D(e, K)$, $z \neq e$, and $w \in \overline{D(e, K)}$, $w \neq e$, the following inclusion holds:

$$F(z) \in \overline{D(\eta, \frac{1+a}{2}\alpha K)}, \qquad (2.4.15)$$

where

$$a = \frac{b+c}{1+bc}, \quad b = \frac{|F'(w)|(1 - |w|^2)}{1 - |F(w)|^2}, \quad c = \left| \frac{z - w}{1 - \bar{w}z} \right|.$$

Note that it follows by Corollary 1.1.1 that $b \leq 1$. Moreover, the Schwarz–Pick Lemma implies that b (hence a) is equal to 1 if and only if F is an automorphism of Δ. If this is not the case (i.e., F is not an automorphism of Δ), then a in (2.2.15) is strictly less than 1, whence this inclusion is stronger that the original one in Julia's Lemma.

Chapter 3

Generation theory on the unit disk

For physical, chemical, and biological applications it is sometimes preferable to study a great variety of iterative processes, including the processes of continuous time. In spite of their simplicity these processes have been applicable in many fields, involving mathematical areas such as geometry, theory of stochastic branching processes, operator theory on Hardy spaces, and optimizations methods. A problem that has interested mathematicians since the time of Abel is how to define n-th iterate of function when n is not integer.

3.1 One-parameter continuous semigroup of holomorphic and ρ-nonexpansive self-mappings

Let A be a topological Abelian (additive) semigroup with zero and let there exist a natural ordering of A, i.e., $\tau \geq t$ if and only if there is $s \in A$ such that $\tau = t + s$.

A mapping $S : A \mapsto \mathrm{Hol}(\Delta)$ (respectively, $N_\rho(\Delta)$) which preserves the additive structure of A with respect to composition operation on $\mathrm{Hol}(\Delta)$ (respectively, $N_\rho(\Delta)$), i.e.,

(i) $S(t + s) = S(t) \circ S(s)$,

whenever s, t and $s + t$ belong to A;

(ii) $S(0) = I$, the identity embedding of Δ, will be referred to its holomorphic (respectively, ρ-nonexpansive) action on Δ.

If $A = \mathbb{N} \bigcup \{0\} = \{0, 1, 2, \ldots\}$ then $S = \{F_0, F_1, F_2, \ldots, F_n, \ldots\}$, $F_n \in \mathrm{Hol}(\Delta)$, (respectively, $N_\rho(\Delta)$) is called a one-parameter discrete semigroup. Actually such a semigroup consists of iterates of a holomorphic (respectively, ρ-nonexpansive) self-mapping $F = F_1$, because of conditions (i) and (ii), i.e., $F_0 = I$, $F_n = F_1^{(n)}$, $n = 1, 2, \ldots$.

If A is an interval of $\mathbb{R}$, containing zero and action $S : A \mapsto \mathrm{Hol}(\Delta)$ $(N_\rho(\Delta))$ is continuous with respect to the topology of pointwise convergence on Δ, we will say that S is *a one-parameter continuous semigroup of holomorphic (ρ-nonexpansive) self-mappings F_t, $t \in A$.*

In other words, a one-parameter continuous semigroup of holomorphic self-mappings of Δ is a family $S = \{F_t\}_{t \in A} \subset \mathrm{Hol}(\Delta)$ $(N_\rho(\Delta))$ such that

(i)' $F_{t+s}(z) = F_t(F_s(z))$, $z \in \Delta$, t, s and $t + s \in A$,
(ii)' $F_0(z) = z$ for all $z \in \Delta$ and

$$\lim_{t \to s} F_t(z) = F_s(z), \qquad (3.1.1)$$

whenever $t, s \in A$.

Mainly we will concentrate on two cases:

(a) $A = [0, T)$, $T > 0$;
(b) $A = (-T, T)$, $T > 0$.

If $T = \infty$ we will say that $S = \{F_t\}_{t \geq 0}$ is a flow on Δ.

In the case (b) conditions (i)' and (ii)' imply that $F_t \circ F_{-t} = F_0 = I$, hence $F_t \in \mathrm{Aut}(\Delta)$ $(\mathrm{Isom}(\Delta))$ and S is actually a group which will be called *a one-parameter group of holomorphic automorphisms (ρ-isometries) of Δ.*

Exercise 1. Prove the following: if $A = [0, T)$ and at least one element F_t of S belongs to $\mathrm{Aut}(\Delta)$, then so does each element of S, and S can be extended to a one-parameter group:

$$\widetilde{S} = \{F_t\}_{t \in (-T, T)} \subset \mathrm{Aut}(\Delta).$$

Example 1. Let a be a complex number such that $\mathrm{Re}\, a \geq 0$ and let $t \in [0, T)$, $T > 0$. Define $F_t \in \mathrm{Hol}(\Delta)$ as follows:

$$F_t(z) = e^{-at} \cdot z, \qquad z \in \Delta. \qquad (3.1.2)$$

It is clear that

$$F_0(z) = z, \qquad z \in \Delta$$

and

$$F_{t+s}(z) = F_t(F_s(z)), \qquad z \in \Delta,$$

whenever $0 \leq s, t < T$ and $t + s < T$.

Since $T > 0$ is arbitrary we have that such a semigroup can be continuously extended to a flow on Δ.

As a matter of fact, we will show below that each one-parameter continuous semigroup of holomorphic self-mappings of Δ can be continuously extended to a holomorphic flow on Δ.

Note also that the semigroup (flow) considered in this example is a linear action on Δ. In fact, each one-parameter semigroup which is a linear action on Δ has the form (3.1.2) with some $a \in \mathbb{C}$, $\operatorname{Re} a \geq 0$.

Example 2. Define the family $\{F_t\} \subset \operatorname{Hol}(\Delta)$, $t \in \mathbb{R}$, by the formula:

$$F_t = \frac{z + \tanh t}{z \tanh t + 1}.$$

It is easy to see that for each $t \in \mathbb{R}$, $F_t \in \operatorname{Aut}(\Delta)$, and that conditions (i) and (ii) are satisfied. Hence, F_t is a flow of automorphisms on Δ.

Exercise 2. Show that each of the automorphisms in Example 2 is hyperbolic.

Example 3. Set

$$F_t(z) = \frac{1 + it}{1 - it} \cdot \frac{z + \frac{t}{1+it}}{1 + \frac{t}{1-it}z}.$$

Since $F_t(z) = \lambda_t m_{a_t}(z)$, where $\lambda_t = \dfrac{1 + it}{1 - it}$ has modulus 1 and $a_t = \dfrac{t}{1 + it} \in \Delta$ for all $t \in \mathbb{R}$, we have that F_t belongs to $\operatorname{Aut}(\Delta)$.

Exercise 3. Let $S = \{F_t\}_{t \in \mathbb{R}}$ be defined as in Example 3. Show that the following statements hold:

(i) $S = \{F_t\}_{t \in \mathbb{R}}$ is a flow of automorphisms of Δ;
(ii) for each $t \neq 0$ the mapping F_t in Example 3 is a parabolic automorphism.

Exercise 4. Consider the family $S = \{F_t\}$ defined as follows:

$$F_t(z) = \frac{z}{e^t - z(e^t - 1)}, \qquad t > 0.$$

Show that (i) for each $t \geq 0$, $|F_t(z)| < 1$, i.e., $F_t \in \operatorname{Hol}(\Delta)$;

(ii) the family $S = \{F_t\}$ is a one-parameter continuous semigroup on Δ and every element of S is not an automorphism of Δ;

(iii) for each F_t, $t > 0$, only elements $\{0, 1\}$ are solutions of the equation

$$F_t(z) = z;$$

(iv) find $\lim_{t \to \infty} F_t(z)$, $z \in \Delta$.

Exercise 5. ([115]) Consider the family $S = \{F_t\}$, where

$$F_t(z) = 1 - \left[1 - e^{-\frac{1}{2}t} + e^{-\frac{1}{2}t}\sqrt{1 - z}\right]^2.$$

(i) Show that $F_t \in \operatorname{Hol}(\Delta)$, and $S = \{F_t\}$ is a one-parameter semigroup;
(ii) Find $\lim_{t \to \infty} F_t(z)$.

Remark 3.1.1 The semigroups in Examples 2 and 3 consist of automorphisms of Δ and consequently are one-parameter groups, while the semigroups in Exercises

4 and 5 cannot be extended to subgroups of Aut(Δ). In fact, F_t in Exercise 4 is defined also for $t < 0$, but it does not map the unit disk Δ into itself. Therefore for $t > 0$ the mapping F_t is biholomorphic on Δ, but $F_t(\Delta)$ is a proper subset of Δ. This fact holds in general.

Proposition 3.1.1 *Let the family* $S = \{F_t\}_{t \in [0,T)}$, $T > 0$ *be a one-parameter continuous semigroup on* Δ, $S \subset \text{Hol}(\Delta)$. *Then each element* $F_t \in S$ *is a local biholomorphism on* Δ.

Proof. Let U be any convex open subset strictly inside Δ. Since S is a normal family on Δ the net $\{F_t\}$ converges to the identity uniformly on $\overline{U}$, as t goes to zero. Therefore for $\delta = \text{dist}(\overline{U}, \partial\Delta)$ and $\varepsilon \in (0, 1 - \delta)$ one can find $t_0 \in (0, T)$ such that $|z - F_t(z)| < \varepsilon$ for all $z \in \overline{U}$ and $t \in [0, t_0)$. It follows by the Cauchy inequality that

$$|1 - (F_t)'(z)| < \frac{\varepsilon}{1 - \delta} < 1,$$

hence $(F_t)'(z) \neq 0$ for all $z \in U$.

Consequently, for each pair z and w in U we have by the Lagrange formula:

$$|F_t(z) - F_t(w)| = |(F_t)'(z + \theta(w - z))| \cdot |z - w|,$$

$0 \leq \theta \leq 1$, equals to zero if and only if $z = w$. So, F_t is injective (hence biholomorphic) on U for all $t \in (0, T)$ small enough.

Now it follows by continuity that if F_{t_0} is locally injective on Δ then so is F_t for all t close enough to t_0. At the same time, if $\{F_t\}$ is a net of locally injective mappings which converges to F_s when t goes to $s \in (0, T)$, then F_s must be either also locally injective or zero. But the latter is impossible. Hence the set of $t \in (0, T)$ such that F_t is locally injective is a nonempty open and closed subset of the interval $[0, T)$. Therefore it must be $[0, T)$. $\square$

3.2   Infinitesimal generator of a one-parameter continuous semigroup

As we mentioned above each one-parameter continuous semigroup of holomorphic self-mappings of Δ defined on an interval $[0, T)$ can be extended to the flow on Δ, i.e., we may always assume that T equals to ∞. Moreover, only the right continuity at zero of a semigroup implies continuity (right and left) on all of $\mathbb{R}^+ = [0, \infty)$. These facts can be shown by different approaches, but here we will establish them by using a very strong property of a continuous one-parameter semigroup of holomorphic self-mappings of Δ to be differentiable with respect to a parameter at each point t on the interval of definition. This nice result for the one-dimensional case is due to E. Berkson and H. Porta [17]. We will give here another proof which can easily be extended to a higher dimension (see, for example, [115, 118], cf., also [5]).

Proposition 3.2.1 *Let $S = \{F_t,\ t \in [0, T)\}$ be a one-parameter semigroup of holomorphic self-mappings of Δ, such that for each $z \in \Delta$:*

$$\lim_{t \to 0^+} F_t(z) = z. \tag{3.2.1}$$

Then for each $z \in \Delta$ there exists the limit:

$$\lim_{t \to 0^+} \frac{z - F_t(z)}{t} = f(z), \tag{3.2.2}$$

which is a holomorphic mapping of Δ into $\mathbb{C}$. Moreover, the convergence in (3.2.2) is uniform on each subset strictly inside Δ.

In other words, if a semigroup $S = \{F_t\}$, $t \in [0, T)$, is right continuous at zero, it is also right-differentiable at zero.

Proof. Step 1. Let $\phi \in \text{Hol}(\Delta)$ and $\{\phi^{(k)}\}$ be its iteration family. Let Δ_r denote the disk centered at zero with radius r. Suppose that there are positive $r_1 < r_2 < 1$ and $0 < \mu < r_2 - r_1$ and an integer $p \geq 1$ such that:

$$\left| z - \phi^{(k)}(z) \right| < \mu \tag{3.2.3}$$

for all $k = 1, 2, \ldots, p$ and $z \in \Delta_{r_2}$.

Thus for all $z \in \Delta_{r_1}$ the following inequality holds

$$\left| z - \phi^{(p)}(z) - p(z - \phi(z)) \right| \leq \frac{\mu}{r_2 - r_1 - \mu}(p - 1) \cdot |z - \phi(z)|. \tag{3.2.4}$$

Indeed, let $z \in \Delta_{r_1}$ and $w \in \Delta_{r_2}$ be such that $|z - w| \leq \mu$. Then the disk $\Delta_{r_2 - r_1 - \mu}(z)$ centered at z with radius $r_2 - r_1 - \mu$ lies in Δ_{r_2}. Hence it follows from (3.2.3) and the Cauchy inequalities that:

$$\left| 1 - (\phi^{(k)})'(z) \right| \leq \frac{\mu}{r_2 - r_1 - \mu}. \tag{3.2.5}$$

Therefore, for $z \in \Delta_{r_1}$ and $w \in \Delta_{r_2}$ such that $|z - w| \leq \mu$ we have by (3.2.5):

$$\left| z - \phi^{(k)}(z) - (w - \phi^{(k)}(w)) \right| < \frac{\mu}{r_2 - r_1 - \mu}|z - w|. \tag{3.2.6}$$

Setting $w = \phi(z)$ and using the triangle inequality we obtain:

$$\begin{aligned}
\left| z - \phi^{(p)}(z) - p(z - \phi(z)) \right| &= \left| \sum_{k=0}^{p-1} \left[\phi^{(k)}(z) - \phi^{(k+1)}(z) - (z - \phi(z)) \right] \right| \\
&\leq \sum_{k=1}^{p-1} \left| \phi^{(k)}(z) - z - \left[\phi^{(k)}(\phi(z)) - \phi(z) \right] \right| \\
&\leq \frac{\mu}{r_2 - r_1 - \mu}(p - 1)|\phi(z) - z|,
\end{aligned}$$

and we are done.

Step 2. Let a semigroup $S = \{F_t\}$, $t \in [0, T)$ satisfy condition (1). For $s > 0$ define the mapping

$$f_s = \frac{1}{s}(I - F_s) \in \mathrm{Hol}(\Delta, \mathbb{C}).$$

We claim that the net $\{f_s\}_{s>0}$ is uniformly bounded on each disk Δ_r, with $0 \leq r < 1$.

Indeed, set $r_1 = r$ and choose $r_2 \in (r_1, 1)$. Take any $\mu > 0$, such that $\mu < r_2 - r_1$ and $\dfrac{\mu}{r_2 - r_1 - \mu} < \dfrac{1}{2}$.

By (3.2.1) one can find $\sigma \in (0, T)$ such that for all $\tau \in (0, \sigma)$ and all $z \in \Delta_{r_2}$

$$|z - F_\tau(z)| < \mu. \tag{3.2.7}$$

Setting $p = [\sigma/s]$ we see that for all $s \in (0, \sigma/2)$, the following relations hold: $p \geq 2$ and $ps \geq \sigma/2$. Also $ks \leq \sigma$ for all $k = 1, 2, \ldots, p$. Hence by (3.2.7) and the semigroup property for all $s \in (0, \sigma/2)$ we have:

$$|z - F_s^{(k)}(z)| < \mu,$$

whenever $z \in \Delta_{r_2}$ and $k = 1, 2, \ldots, p(= p(s))$.

From Step 1 we obtain for all $z \in \Delta_r$:

$$
\begin{aligned}
p|z - F_s(z)| - |z - F_s^{(p)}(z)| &\leq |p(z - F_s(z)) - (z - F_s^{(p)}(z)| \\
&\leq \frac{1}{2} p|z - F_s(z)|
\end{aligned}
$$

or

$$|z - F_s(z)| < \frac{2}{p}\left|z - F_s^{(p)}(z)\right|, \tag{3.2.8}$$

whenever $s \in (0, \sigma/2)$ and $p = [\sigma/s]$.

So, by (3.2.7) and (3.2.8) we obtain:

$$|f_s(z)| \leq \frac{2}{ps}|z - F_{sp}(z)| < \frac{2\mu}{ps} \leq \frac{4\mu}{\sigma} < \infty,$$

as claimed.

Step 3. Now we will show that the net $\{f_s\}$, $f_s = 1/s(I - F_s)$, $s \in (0, T)$, converges to a holomorphic mapping f on Δ, when s goes to 0^+. Set $n = n(s) = [1/s^2]$ and consider the sequence $\{f_{1/n}\}$. Since by Step 2 this sequence is uniformly bounded for all large n one can find a subsequence $\{f_{1/n_k}\}$ which converges to a mapping $f \in \mathrm{Hol}(\Delta, \mathbb{C})$ uniformly on each compact subset of Δ. That is for each $r \in (0, 1)$ and all $z \in \Delta_r$ and for a given $\varepsilon > 0$ we can choose k large enough such that:

$$\left|f_{1/n_k}(z) - f(z)\right| < \varepsilon. \tag{3.2.9}$$

In addition, for such n_k and $s \in (0, 1)$ we have:

$$
\begin{aligned}
f_s - f_{\frac{1}{n_k}} &= \frac{1}{s}(I - F_s) - n_k(I - F_{\frac{1}{n_k}}) \\
&= \frac{1}{s}\left(F_{\frac{1}{n_k}}^{([sn_k])} - F_s\right) + \frac{1}{s}\left[(I - F_{\frac{1}{n_k}}^{([sn_k])}) - n_k s(I - F_{\frac{1}{n_k}})\right].
\end{aligned}
$$

Observe, that since, $n = \left[1/s^2\right]$, we have $ns \to \infty$ and $[ns]/ns \to 1$ as $s \to 0^+$. Moreover, for a given net $s \to 0^+$ we can find (if necessary) a sequence $s_k \to 0^+$ such that $s_k/s \le 1$.

Then, setting $n_k = \left[1/s_k^2\right]$, we have:

$$\frac{s}{s_k^2} - s < sn_k = s\left[\frac{1}{s_k^2}\right] < \frac{s}{s_k^2} + 1 \cdot s.$$

Since, $s/s_k \ge 1$ we obtain:

$$\frac{s}{s_k^2} \ge \frac{1}{s_k} \to \infty \quad \text{as} \quad s_k \to 0^+.$$

Therefore $n_k s \to \infty$ as $s \to 0^+$ and $[n_k s]/n_k s \to 1$ as $s \to 0^+$. Thus we can find $\delta > 0$ such that $1 - [n_k s]/n_k s < \varepsilon$ and $F_{[n_k s]/n_k}(z) \subset \Delta_{r_2} \subset \Delta$, whenever $s \in (0,\delta)$ and $z \in \Delta_r$, $r < r_2 < 1$. Then we obtain by step 1 and the semigroup property:

$$
\begin{aligned}
\frac{1}{s}\left|F_{\frac{[n_k s]}{n_k}}(z) - F_s(z)\right| &= \frac{1}{s}\left|F_{\frac{[n_k s]}{n_k}}(z) - F_{\frac{[n_k s]}{n_k}} \circ F_{s-\frac{[n_k s]}{n_k}}(z)\right| \\
&\le \frac{1}{s}M\left|z - F_{s-\frac{[n_k s]}{n_k}}(z)\right| \\
&< \frac{4\mu}{\sigma s}\left(s - \frac{[n_k s]}{n}\right),
\end{aligned}
\tag{3.2.10}
$$

where $M = \sup|(F_s)'(z)|$.

Finally we obtain by Step 1:

$$
\begin{aligned}
\frac{1}{s}&\left|z - F_{\frac{1}{n_k}}^{([sn_k])}(z) - n_k s\left(z - F_{\frac{1}{n_k}}(z)\right)\right| \\
&\le \frac{1}{s}\left|z - F_{\frac{1}{n_k}}^{([sn_k])}(z) - [n_k s]\left(z - F_{\frac{1}{n_k}}(z)\right)\right| + \frac{1}{s}\left|[n_k s] - n_k s\right| \cdot \left|z - F_{\frac{1}{n_k}}(z)\right| \\
&\le \frac{\varepsilon}{s}\left([n_k s] + |[n_k s - ns]|\right)\left|z - F_{\frac{1}{n_k}}(z)\right| \\
&\le \left(\varepsilon\frac{[n_k s]}{n_k s} + \left|\frac{[n_k s]}{n_k s} - 1\right|\right)n_k|z - F_{\frac{1}{n_k}}(z)| \\
&\le \frac{8\mu}{\sigma}\varepsilon.
\end{aligned}
\tag{3.2.11}
$$

Thus for $z \in \Delta_r$ and $s \in (0,\delta)$ we obtain from (3.2.10) and (3.2.11):

$$
\begin{aligned}
|f_s(z) - f(z)| &\le \left|f_{\frac{1}{n_k}}(z) - f(z)\right| + \left|f_s(z) - f_{\frac{1}{n_k}}(z)\right| \\
&\le \frac{16\mu}{\sigma}\varepsilon
\end{aligned}
$$

and we have completed the proof. $\square$

Further we will show that even for the wider class of semigroups of the ρ-nonexpansive mappings the property of the right-differentiability at zero, implies the continuous extension of the semigroup to a flow in $N_\rho(\Delta)$ and the differentiability of this flow at each point t on $(0, \infty)$.

Definition 3.2.1 *Let* $S = \{F_t\} \subset N_\rho(\Delta)$, $t \in [0, T)$, $T > 0$, *be a one-parameter semigroup of* ρ-*nonexpansive self-mappings of* Δ. *If there exists the pointwise limit* $f : \Delta \mapsto \mathbb{C}$ *defined by*

$$f = \lim_{t \to 0^+} \frac{I - F_t}{t}, \tag{3.2.12}$$

we will say that S *is generated by* f *and* f *is called the infinitesimal generator of the semigroup* $S = \{F_t\}_{t \in [0, T)}$.

If this will not imply a confusion we will often simply say 'generator of a semigroup', omitting the word 'infinitesimal'.

Thus Proposition 3.2.1 states the existence and holomorphity of generators for every one-parameter continuous (even only right-continuous at 0^+) semigroup of holomorphic self-mapping of Δ.

Proposition 3.2.2 *Let* $S = \{F_t\}_{t \in [0, T)}$ *be a one-parameter semigroup of* ρ-*nonexpansi self-mappings on* Δ, *which is generated by* $f : \Delta \mapsto \mathbb{C}$ *and suppose that its generator* f, *defined by (3.2.12), is a continuous function on* Δ. *Then the complex valued function* $u(\cdot, \cdot)$ *defined on* $[0, T) \times \Delta$ *by* $u(t, z) = F_t(z)$ *is the solution of the following Cauchy problem:*

$$\begin{cases} \dfrac{\partial u(t, z)}{\partial t} + f(u(t, z)) = 0, & t \in [0, T), \\[3mm] u(0, z) = z, & z \in \Delta. \end{cases} \tag{$*$}$$

Proof. Fix $s \in [0, T)$ and set $F_s(z) = w$. We have:

$$\begin{aligned} \frac{\partial^+ u(t, z)}{\partial t}\bigg|_{t=s} &= \lim_{h \to 0^+} \frac{F_{s+h}(z) - F_s(z)}{h} = \lim_{h \to 0^+} \frac{F_h(F_s(z)) - F_s(z)}{h} \\[2mm] &= \lim_{h \to 0^+} \frac{F_h(w) - w}{h} = -f(w) = -f(u(s, z)). \end{aligned}$$

This means that the right-hand partial derivative of $u(t, z) = F_t(z)$ in $t = s \in [0, T)$ exists and is exactly equal to $-f(u(s, z))$.

Now for fixed $s \in (0, T)$ we denote $w_h = F_{s-h}(z)$ for $h \in [0, \varepsilon)$ with ε small enough. Since z is an interior point of Δ we can find ε small enough such that the set $\{w_h\}_{h \in [0, \varepsilon)}$ lies in a compact subset Ω of Δ. Then we obtain by the uniform continuity of f on Ω

$$\begin{aligned} \frac{\partial^- u(t, z)}{\partial t}\bigg|_{t=s} &= -\lim_{h \to 0^+} \frac{F_{s-h}(z) - F_s(z)}{h} \\[2mm] &= -\lim_{h \to 0^+} \frac{w_h - F(w_h)}{h} \\[2mm] &= -\lim_{h \to 0^+} f_h(w_h) = -f(w), \end{aligned}$$

where
$$w = \lim_{h \to 0^+} w_h = F_s(z).$$

So $u(t, z)$ is also left-differentiable on the interval $(0, T)$ and therefore satisfies the Cauchy problem (*). □

Remark 3.2.1 As a matter of fact, it is known (see, for example, [159]) that if a function has a right-hand derivative which is continuous, then it also has a left-hand derivative and they coincide.

Corollary 3.2.1 *If $S = \{F_t\}$, $t \in [0, T)$ is a one-parameter semigroup of holomorphic self-mappings, which is right-continuous at zero, then it is continuous at each point of $[0, T)$ and, moreover, it is differentiable on this interval.*

Remark 3.2.2 Also we note that since the generator f of a semigroup of holomorphic mappings is holomorphic, it is locally Lipschitzian. Therefore the Cauchy problem (*) has a unique local solution in a neighborhood of $t = 0$ for each initial value $z \in \Delta$. This fact and the analytic continuation principle can be successfully used to prove the extension of the semigroup to a flow of holomorphic mappings on Δ (see [115]).

However, to prove this for the class of ρ-nonexpansive mappings we will use another approach which is based on the so called resolvent method. We will discuss it in the next section.

At the end of this section we present a property which is specific for semigroups of holomorphic self-mappings.

Exercise 1. Prove that if $u(\cdot, \cdot)$ is a jointly continuous function on $[0, T) \times \Delta$ which satisfies (*) with $f \in \text{Hol}(\Delta, \mathbb{C})$, then it also satisfies the linear partial differential equation:
$$\frac{\partial u(t, z)}{\partial t} + \frac{\partial u(t, z)}{\partial z} f(z) = 0.$$

Exercise 2. Find the (infinitesimal) generators in Example 1–5 of Section 3.1.

3.3 Nonlinear resolvent and the exponential formula

In this section we consider the notion of the nonlinear resolvent of a continuous (or holomorphic) function and establish a result in the spirit of the Hille–Yosida theory (see, for example, [159]). More precisely, we will show that the global solvability of the Cauchy problem (*) (see Section 3.2) is equivalent to the solvability of a functional equation (see [115, 116, 118]). To make the ideas more transparent we first explain them for the simplest linear case.

We already mentioned that the only linear (in the complex sense) semigroup S on $\mathbb{C}$ is of the form: $S = \{F_t\}_{t>0}$, $F_t(z) = \exp(-ta)z$, $a \in \mathbb{C}$, with the infinitesimal generator $f : \mathbb{C} \mapsto \mathbb{C}$ defined by

$$f(z) := \lim_{t \to 0^+} \frac{z - F_t(z)}{t} = az.$$

Conversely, for a given number $a \in \mathbb{C}$ two following equivalent definitions of the exponential function

$$u(t, z) = \exp(-ta) z$$

usually are employed in the classical analysis.

The first one is based on the solution of the linear Cauchy problem:

$$\begin{cases} \dfrac{\partial u(t, z)}{\partial t} + au(t, z) = 0 \\[2mm] u(0, z) = z \in \Delta, \end{cases} \tag{3.3.1}$$

$a \in \mathbb{C}$, while the second one uses the exponential formula:

$$u(t, z) = \lim_{n \to \infty} \left(1 + \frac{t}{n} a\right)^{-n} z. \tag{3.3.2}$$

On the other hand, if for given $z \in \mathbb{C}$, $a \in \mathbb{C}$, and $r \geq 0$, we solve the linear equation

$$w + raw = z, \tag{3.3.3}$$

we have that its solution

$$w = J_r(z) = (1 + ra)^{-1} z$$

can be considered as a linear mapping defined on $\mathbb{C}$. Then the exponential formula (3.3.2) can be rewritten in the form

$$u(t, z) = \lim_{n \to \infty} J_{t/n}^{(n)}(z), \tag{3.3.4}$$

where $J_r^{(n)}$ denotes the n-fold iterate of the mapping J_r, $r \geq 0$.

In addition, it is clear that $S = \{\exp(-ta)\}$ is a ρ-nonexpansive action on the unit disk Δ if and only if

$$\operatorname{Re} a \geq 0.$$

This is, in fact, equivalent to the property that for each $r \geq 0$ the mapping $J_r : \mathbb{C} \mapsto \mathbb{C}$ is a self-mapping of Δ, hence is ρ-nonexpansive.

Now, if we consider, for example, a (nonlinear) analytic function $f : \Delta \mapsto \mathbb{C}$ defined by

$$f(z) = z - z^2,$$

we see that the solution of the Cauchy problem

$$\begin{cases} \dfrac{\partial u(t, z)}{\partial t} + f\left(u(t, z)\right) = 0, \\ u(0, z) = z \in \Delta, \end{cases} \tag{3.3.5}$$

is a well defined semigroup (even a flow) of holomorphic (hence ρ-nonexpansive) mappings of Δ:

$$u(t, z)(= F_t(z)) = \frac{z}{e^t - z(e^t - 1)}$$

(see Exercise 4 in Section 3.1).

But in this case the similar formula (3.3.2) makes no sense with respect to (3.3.5).

At the same time, it turns out, that formula (3.3.4) continues to hold with $J_r : \Delta \mapsto \Delta, \quad r \geq 0$ defined as the solution w of the equation

$$w + r(w - w^2) = z \tag{3.3.6}$$

(compare with equation (3.3.3)).

Exercise 1. Show directly that for each $z \in \Delta$ and $r > 0$ the quadratic equation (3.3.6) has a unique solution $w = J_r(z) \in \Delta$ which holomorphically depends on $z \in \Delta$. Show that $\lim_{n \to \infty} J_{t/n}^{(n)}(z) = F_t(z)$.

In general such an approach, which we call the resolvent method, is very useful in the study of generated ρ-nonexpansive semigroups and their properties.

In addition, by using this method one can give different characterizations of the class of functions which generate a flow of ρ-nonexpansive or holomorphic mappings, or in other words, to give sufficient and necessary conditions for the global solvability of the Cauchy problem. To continue we need the following definition.

Definition 3.3.1 *Let* $f : \Delta \mapsto \mathbb{C}$ *be a continuous function. We will say that* f *satisfies the range condition (RC) if for each* $r > 0$ *the nonlinear resolvent* $J_r = (I + rf)^{-1}$ *is well defined on* Δ *and belongs to* $N_\rho(\Delta)$.

In other words, f satisfies the range condition (RC) if for each $r > 0$ and $z \in \Delta$ the equation

$$w + rf(w) = z \tag{3.3.7}$$

has a unique solution $w = J_r(z)$ in Δ, such that

$$\rho(J_r(w_1), \, J_r(w_2)) \leq \rho(w_1, w_2),$$

whenever $w_1, \, w_2$ belong to Δ.

Proposition 3.3.1 *A continuous (holomorphic) function f on Δ is an infinitesimal generator of a one-parameter semigroup $S = \{F_t\} \subset N_\rho(\Delta)$ (respectively,
$\mathrm{Hol}(\Delta)$) defined on $[0,T)$, $T > 0$ if and only if it satisfies the range condition
(RC).*

*Moreover, the semigroup $S = \{F_t\}$ is unique and can be continuously extended
to a flow of ρ-nonexpansive (holomorphic) self-mappings defined on $\mathbb{R}^+ = [0,\infty)$
by the following exponential formula*

$$F_t(z) = \lim_{n\to\infty} J_{t/n}^{(n)}(z) \quad z \in \Delta, \ t > 0. \tag{3.3.8}$$

Since the proof of this assertion is rather long we will give it step by step using
several lemmata which will also be needed independently in the sequel.

Lemma 3.3.1 *Let ρ be the Poincaré metric on Δ and let $\{G_t\}$, $t \in [0,T)$, $T > 0$
be a family of ρ-nonexpansive mappings of Δ, i.e.,*

$$\rho(G_t(z), \ G_t(w)) \leq \rho(z,w) \tag{3.3.9}$$

for all $z, w \in \Delta$, $t \in [0,T)$. Suppose that for each $z \in \Delta$ there exists the limit:

$$f(z) = \lim_{t\to 0^+} \frac{1}{t}(z - G_t(z)) \tag{3.3.10}$$

*and assume that f is continuous on each compact subset of Δ. Then for each
$r > 0$ and each $w \in \Delta$ the equation*

$$w + rf(w) = z \tag{3.3.11}$$

has a unique solution $z = J_r(w)$ and $J_r : \Delta \mapsto \Delta$ is also ρ-nonexpansive.

Proof. Given $t \in (0,T)$ denote:

$$f_t = \frac{1}{t}(I - G_t) \tag{3.3.12}$$

and consider the equation:

$$w + rf_t(w) = z, \quad z \in \Delta, \quad r > 0. \tag{3.3.13}$$

This equation can be rewritten in the form:

$$w = \frac{r}{r+t}G_t(w) + \frac{t}{r+t}z. \tag{3.3.14}$$

For fixed $z \in \Delta$ the mapping defined by the right-hand side of (3.3.14) is a
strict contraction with respect to the metric ρ (because of (3.3.9) and Proposition
2.3.1(ii)). Therefore, for each $z \in \Delta$ and $r \geq 0$, this equation has a unique
solution $w_t = J_{r,t}(z) \in \Delta$. In addition, this solution can be obtained by the
iteration method:

$$w_{n+1}(= w_{n+1}(z)) = \frac{r}{r+t}G_t(w_n) + \frac{t}{r+t}z, \tag{3.3.15}$$

where z_0 is an arbitrary element in Δ.

Setting $w_0(z) = z$ we have by induction and Proposition 2.3.1(iii) that

$$\begin{aligned}
\rho(w_{n+1}(z_1), w_{n+1}(z_2)) &\leq \max\{\rho(G_t(w_n(z_1)), G_t(w_n(z_2))), \rho(z_1, z_2)\} \\
&\leq \max\{\rho(w_n(z_1), w_n(z_2)), \rho(z_1, z_2)\} \leq \rho(z_1, z_2).
\end{aligned}$$

It means that all $w_n(\cdot)$ are ρ-nonexpansive and so is $J_{r,t}(\cdot) = \lim_{n\to\infty} w_n(\cdot)$. Note, in passing, that if $G_t \in \mathrm{Hol}(\Delta)$ then $J_{r,t} : \Delta \mapsto \Delta$ defined as the solution of (3.3.14) is also holomorphic on Δ.

Now we want to show that for some $r > 0$ and $z \in \Delta$ the net $\{J_{r,t}(z)\}_{t\in(0,T)}$ converges to $J_r(z)$, as t tends to 0^+ and its limit is a solution of equation (3.3.11).

To do this we first claim that this net lies strictly inside Δ. The latter is equivalent to the inequality:

$$\rho(z, J_{r,t}(z)) \leq M \ (= M(z)) < \infty, \text{ as } t \to 0^+. \tag{3.3.16}$$

Indeed, since

$$J_{r,t}(z) = \frac{r}{r+t} G_t(J_{r,t}(z)) + \frac{t}{r+t} z,$$

we have by Proposition 3.3.1(ii):

$$\begin{aligned}
\rho\Big(J_{r,t}(z) &, \ \frac{r}{r+t} z + \frac{t}{r+t} z \Big) \\
&\leq \ \Big(\frac{r}{r+t} + \frac{t}{r+t} |z| \Big) \rho(G_t(J_{r,t}(z)), z) \\
&\leq \ \Big(\frac{r}{r+t} + \frac{t}{r+t} |z| \Big) [\rho(G_t(J_{r,t}(z)), G_t(z)) + \rho(G_t(z), z))] \\
&\leq \ \Big(\frac{r}{r+t} + \frac{t}{r+t} |z| \Big) [\rho(J_{r,t}(z), z) + \rho(G_t(z), z)] .
\end{aligned}$$

This inequality implies that

$$\rho(J_{r,t}(z), z) \leq \frac{r+t}{t(1-|z|)} \Big(\frac{r}{r+t} + \frac{t}{r+t} |z| \Big) \rho(G_t(z), z).$$

Since for $t \in (0, T)$ small enough the element $G_t(z)$ is close to z we have that there exists a positive number $M_1 < \infty$ such that

$$\begin{aligned}
\limsup_{t\to 0^+} \frac{1}{t} \rho(G_t(z), z) &\leq \ \limsup_{t\to 0^+} \frac{M_1}{t} |G_t(z) - z| \\
&= \ M_1 |f(z)|.
\end{aligned}$$

Consequently, we have the estimate:

$$\limsup_{t\to 0^+} \rho(J_{r,t}(z), z) \leq \frac{r}{1-|z|} M_1 |f(z)| =: M,$$

which implies (3.3.16). $\square$

Thus this lemma proves the necessary assertion of Proposition 3.3.1 if we set $G_t = F_t$, $t \in [0, T)$, $T > 0$.

To accomplish this matter we note that it follows by the uniqueness of the local solution of the Cauchy problem (see also remark at the end of this section) that if f is holomorphic in Δ, then so is F_t for each $t \in [0, T)$.

Then our constructions in Lemma 3.3.1 show that the resolvent $J_r : \Delta \to \Delta$ is a holomorphic mapping for each $r > 0$. This fact can be shown also by using the local Implicit Function Theorem (see, for example, [115]).

Let as above ρ be the hyperbolic Poincaré metric on Δ and $B(a, R) = \{w \in \Delta : \rho(a, w) < R\}$, $a \in \Delta$, $R > 0$.

Lemma 3.3.2 *Let f be a continuous function which satisfies the range condition (RC). Then for each $a \in \Delta$ and $R > 0$ there are $\tau = \tau(a, R)$, $0 < \tau < 1$ and $L = L(a, R) < \infty$ such that*

$$\rho(J_r^{(k)}(z), z) \le rkL$$

for all $r \in (0, \tau)$ and $k = 0, 1, 2, \ldots$.

Proof. Since each ρ-ball is bounded away from the boundary of Δ for given $a \in \Delta$ and $R > 0$ we can find $0 < s < 1$ such that $\overline{B(a, R)} \subset \Delta_s = \{z \in \Delta : |z| < s\}$.

Denote $M = \max\{|f(w)|, w \in B(a, R)\}$ and set $\tau = d/M$, where $d = \text{dist}\{\partial B(a, R), \partial \Delta_s\} > 0$.

Then for each $r \in (0, \tau)$ and $w \in B(a, R)$ we have

$$z = w + rf(w) \in \Delta \qquad (3.3.17)$$

and $w = J_r(z) \in B(a, R)$.

Hence, for such r and all $w \in B(a, R)$ we obtain by (3.3.17)

$$\begin{aligned} \rho(J_r(w), w) &= \rho(J_r(w), J_r(z)) \le \rho(z, w) \\ &\le \tanh^{-1} \frac{|z - w|}{1 - s^2} = \text{arctanh} \frac{rM}{1 - s^2}. \end{aligned} \qquad (3.3.18)$$

Further, it follows by the Lagrange mean value theorem that for each $t \in [0, t_0]$, $t_0 < 1$,

$$\tanh^{-1} t \le t \cdot \frac{1}{1 - t_0^2}.$$

Then setting

$$t_0 = \frac{\tau M}{1 - s^2} = \frac{d}{1 - s^2} \text{ and } L = \frac{M(1 - s^2)}{(1 - s^2)^2 - d^2}$$

we obtain by using (3.3.18):

$$\rho(J_r(w), w) \le rL$$

for all $w \in B(a, R)$ and $r \in [0, \tau)$.

Now using the triangle inequality we have

$$\rho(J_r^{(k)}(z), z) \leq \sum_{j=1}^{k} \rho\left(J_r^{(j)}(z), J_r^{(j-1)}(z)\right) \leq k\rho(J_r(z), z).$$

Hence

$$\rho(J_r^{(k)}(z), z) \leq rkL$$

and the Lemma is proved. $\square$

Lemma 3.3.3 *Let f be a continuous function in Δ which satisfies the range condition (RC). Then for each $a \in \Delta$, $R > 0$, and $\varepsilon > 0$, there is $\mu = \mu(a, R, \varepsilon) > 0$ such that for all $r \in [0, \mu)$ and each $p = 0, 1, 2, \ldots$ the following inequalities hold*

$$\left| f(z) - \frac{z - J_{r/p}^{(p)}(z)}{r} \right| \leq \varepsilon \tag{3.3.19}$$

and

$$\left| J_r(z) - J_{r/p}^{(p)}(z) \right| < 2r\varepsilon, \tag{3.3.20}$$

whenever $z \in B(a, R)$.

Proof. Since f is continuous in Δ, for each $a \in \Delta$, $R > 0$ and $\varepsilon > 0$ one can find $\delta > 0$ such that $|f(z) - f(w)| < \varepsilon$, whenever z and w belong to $B(a, R)$ and $\rho(z, w) < \delta$.

Let $\tau = \tau(a, R)$ and $L = L(a, R)$ be found as in Lemma 3.3.2, and set $\mu = \min\{\tau, \delta/L\}$. Then for all $r \in (0, \mu)$ and each $p = 1, 2, \ldots$ we obtain by this Lemma

$$\rho(z, J_{r/p}^{(k)}(z)) \leq kL\frac{r}{p} < \frac{k}{p}\delta, \quad z \in B(a, R).$$

Hence, for all $k = 1, \ldots, p$ we have

$$\left| f(z) - f(J_{r/p}^{(k)}(z)) \right| < \varepsilon, \quad z \in B(a, R).$$

In addition, it follows by the definition of the resolvent that for all $w \in \Delta$

$$J_{r/p}(w) - w = \frac{r}{p} f\left(J_{r/p}(w)\right).$$

Now using the triangle inequality we estimate

$$\left| f(z) - \frac{z - J_{r/p}^{(p)}(z)}{r} \right| = \frac{1}{r}\left| rf(z) - z + J_{r/p}^{(p)}(z) \right|$$

$$= \frac{1}{r}\sum_{k=1}^{p}\left| \frac{r}{p}f(z) - J_{r/p}^{(k-1)}(z) + J_{r/p}^{(k)}(z) \right|$$

$$\leq \frac{1}{r}\sum_{k=1}^{p}\left| \frac{r}{p}f(z) + J_{r/p}(J_{r/p}^{(k-1)}(z)) - J_{r/p}^{(k)}(z) \right|$$

$$= \frac{1}{r}\sum_{k=1}^{p}\frac{r}{p}\left| f(z) - f(J_{r/p}^{(k)}(z)) \right| \leq \varepsilon,$$

whenever $z \in B(a, R)$.

So, (3.3.19) is proved. In turn (3.3.19) implies (3.3.20):

$$\left| J_r(z) - J_{r/p}^{(p)}(z) \right| \leq \left| z - J_{r/p}^{(p)}(z) - rf(z) \right| + |rf(z) - z + J_r(z)|$$

$$\leq r \left| f(z) - \frac{z - J_{r/p}^{(p)}}{r} \right| + r |f(z) - f(J_r(z))| \leq 2r\varepsilon,$$

and we have completed the proof. $\square$

Lemma 3.3.4 (The resolvent identity) *Let f be a continuous function in Δ which satisfies the range condition (RC). Then for $0 \leq s \leq t$ the following resolvent identity holds:*

$$J_t(z) = J_s \left(\frac{s}{t} z + \left(1 - \frac{s}{t} \right) J_t(z) \right), \quad z \in \Delta.$$

Proof. Since for each $z \in \Delta$ and $t \geq 0$ the element $J_t(z) \in \Delta$ we have

$$w = \frac{s}{t} z + \left(1 - \frac{s}{t} \right) J_t(z) \in \Delta \qquad (3.3.21)$$

by the convexity of Δ. It follows by the definition of the resolvent that

$$z - J_t(z) = tf(J_t(z)) \qquad (3.3.22)$$

and

$$J_s(w) + sf(J_s(w)) = w. \qquad (3.3.23)$$

On the other hand (3.3.21) and (3.3.22) imply

$$w = J_t(z) + \frac{s}{t}(z - J_t(z)) = J_t(z) + sf(J_t(z)).$$

Since equation (3.3.23) has a unique solution we have the equality

$$J_t(z) = J_s(w),$$

which is equivalent to the resolvent identity. $\square$

Now we are able to complete the proof of Proposition 3.3.1. As we mentioned above the necessity of the assertion of this Proposition follows from Lemma 3.3.1.

To prove the sufficiency we first show that for all $t \geq 0$ and $z \in \Delta$ the limit in (3.3.8) exists. In fact, it is enough to prove that for each $t \geq 0$ and $z \in \Delta$ the sequence $\{J_{t/n}^{(n)}(z)\}$ is a Cauchy sequence in the Poincaré metric on Δ.

Indeed, fix any $t \geq 0$ $a \in \Delta$ and consider the sequence $\{z_l\}_{l=1}^{\infty} \subset \Delta$ defined as follows

$$z_l = J_{t/l}^{(k(l))}(a), \quad k(l) \leq l. \qquad (3.3.24)$$

For an arbitrary $s \in (0, 1)$ close to 1, one can find l_0 such that elements

$$w_l = a + \frac{t}{l} f(a) \in \Delta_s = \{|z| < s < 1\},$$

whenever $l \geq l$. Also, for such l we have $J_{t/l}(w_l) = a$. Then as in Lemma 3.3.2 we obtain that there is $d = d(s) > 0$ such that for all $l \geq l_0$

$$
\begin{aligned}
\rho\left(J_{t/l}^{(k(l))}(a), a\right) &\leq k(l)\rho\left(J_{t/l}(a), a\right) = k(l)\rho\left(J_{t/l}(a), J_{t/l}(w_l)\right) \\
&\leq k(l)\rho(a, w_l) \leq k(l)|w_l - a| \cdot \frac{d}{1 - s^2} \\
&= \frac{k(l)}{l} t|f(a)| \cdot \frac{d}{1 - s^2} \leq t|f(a)|\frac{d}{1 - s^2}.
\end{aligned}
$$

Setting now

$$
R = \max\left\{\rho(a, z_l), \ l - 1, 2, \ldots, l_0, t|f(a)|\frac{d}{1 - s^2}\right\}
$$

we obtain that the sequence $\{z_l\}_{l=1}^{\infty}$ defined by (3.3.24) is contained in $B(a, R)$.

Using this fact we will show now that the sequence $\{J_{t/n}^{(n)}(a)\}$ is a Cauchy sequence in Poincaré metric on Δ.

Taking any $\varepsilon > 0$ and $z \in B(a, R)$, define $\mu > 0$ as in Lemma 3.3.3. Then we have by this lemma that there are $n_0 > 0$, $m_0 > 0$ and $L > 0$ such that for all $n > n_0$ and $m > m_0$

$$
\rho\left(J_{t/n}(z), J_{t/nm}^{(m)}(z)\right) < \frac{2L\varepsilon t}{n}
$$

and

$$
\rho\left(J_{t/m}(z), J_{t/nm}^{(n)}(z)\right) < \frac{2L\varepsilon t}{n},
$$

whenever $z \in B(a, R)$.

Using these inequalities we obtain after several manipulations with the triangle inequality that

$$
\rho\left(J_{t/n}^{(n)}(a), J_{t/m}^{(m)}(a)\right) \leq 4Lt\varepsilon,
$$

as required.

Since a is an arbitrary element of Δ we have that the limit in (3.3.8) exists and the mapping $F_t : \Delta \mapsto \Delta$ defined by this formula is a ρ-nonexpansive mapping on Δ.

Now we have to establish the continuity of $\{F_t(z)\}_{t \geq 0}$, for each $z \in \Delta$.

By using (3.3.8) it is enough to prove continuity of the resolvent $J_r(z)$ for r sufficiently small.

To this end let us choose $0 < t < r < \tau$ for any $a \in \Delta$ and $R > 0$ as in Lemma 3.3.1. Then we have for such r, that $\rho(J_r(a), a) < R$ and by the resolvent identity (Lemma 3.3.4) we obtain

$$
\begin{aligned}
\rho\left(J_t(a), J_r(a)\right) &= \rho\left(J_t(\frac{t}{r}a + \frac{r-t}{r}J_r(a)), J_t(a)\right) \\
&\leq \rho\left(\frac{t}{r}a + \frac{r-t}{r}J_r(a), a\right) \leq \rho(a, a - (r - t)f(J_r(a)) \to 0
\end{aligned}
$$

as $r - t \to 0$.

The semigroup property of $S = \{F_t\}_{t \geq 0}$ can be proved in a standard way (see, for example, [29]), using (3.3.8) and passing from rational $t \geq 0$, $s \geq 0$ to real numbers by the continuity of $\{F_t\}_{t \geq 0}$.

Now using (3.3.19) (see Lemma 3.3.3) and again (3.3.8) one can easily show that for each $z \in \Delta$

$$\lim_{t \to 0^+} \frac{z - F_t(z)}{t} = f(z).$$

Thus $f : \Delta \mapsto \mathbb{C}$ is the infinitesimal generator of the flow $S = \{F_t\}_{t \geq 0}$, defined by (3.3.8).

Finally, the uniqueness of this flow follows by Proposition 3.2.2, and the proof is complete. $\square$

Remark 3.3.1 By $\mathcal{G}N_\rho(\Delta)$ (respectively, $\mathcal{G}\,\mathrm{Hol}(\Delta)$) we will denote the set of all continuous (respectively, holomorphic) functions on Δ which are generators of one-parameter semigroups (flows) of ρ-nonexpansive (respectively, holomorphic) self-mappings of Δ.

A direct consequence of the above Proposition and Lemma 3.3.1 is that *these sets are real cones, i.e., if f and g belong to $\mathcal{G}N_\rho(\Delta)$ (respectively, $\mathcal{G}\,\mathrm{Hol}(\Delta)$) then so does the function*

$$h = \alpha f + \beta g$$

for each pair of nonnegative numbers α and β.

Indeed, let $\{F_t\}_{t \geq 0}$, $\{G_t\}_{t \geq 0}$ be the flows generated by f and g respectively. Then if we define the family $\{H_t\}_{t \geq 0}$ by

$$H_t(z) = F_{\alpha t}(G_{\beta t}(z))$$

we have that

$$h(z) = \alpha f(z) + \beta g(z) = \lim_{t \to 0^+} \frac{z - H_t(z)}{t}$$

for each $z \in \Delta$, and we are done.

This fact can be also established by using representation theory of generators or the so called flow invariance conditions (see following sections).

Moreover, we will see below that the sets $\mathcal{G}N_\rho(\Delta)$ and $\mathcal{G}\,\mathrm{Hol}(\Delta)$ are closed with respect to the open compact topology on Δ. $\square$

Exercise 1. Prove directly that if $f_n \in \mathcal{G}\,\mathrm{Hol}(\Delta)$ is a convergent sequence on each compact subset of Δ, then its limit function f also belongs to $\mathcal{G}\,\mathrm{Hol}(\Delta)$.

Hint: Use Proposition 3.2.2 and the properties of the solution of the Cauchy problem (*) (see section 3.2).

At the end of this section we will give another important consequence of Proposition 3.3.1, which will be used in the sequel.

Corollary 3.3.1 *Let $F : \Delta \mapsto \Delta$ be a ρ-nonexpansive (respectively, holomorphic) self-mapping of Δ. Then $f = I - F$ (i.e., $f(z) = z - F(z)$) belongs to $\mathcal{G}N_\rho(\Delta)$ (respectively, $\mathcal{G}\,\mathrm{Hol}(\Delta)$).*

Proof. We have to show that for each $r \geq 0$ and $z \in \Delta$ the equation

$$w + r(w - F(w)) = z \tag{3.3.25}$$

has a unique solution $w = J_r(z)$ in Δ.

Indeed, setting $t = r/(r+1)$ we can rewrite (3.3.25) in the form

$$w = tF(w) + (1-t)z. \tag{3.3.26}$$

It was shown in Section 2.4 that for each $t \in [0,1)$ equation (3.3.26) has a unique solution $w = G_t(z)$ and that for each $t \in [0,1)$ the mapping $G_t : \Delta \mapsto \Delta$ is a ρ-nonexpansive self-mapping of Δ (see Proposition 2.4.1). In addition, if $F \in \text{Hol}(\Delta)$, then so is G_t (Remark 2.4.1). Thus the function $f = I - F$) satisfies the range condition and its resolvent $J_r : \Delta \mapsto \Delta$, $r \geq 0$ is defined by

$$J_r(z) = G_{r/(1-r)}(z). \tag{3.3.27}$$

This completes the proof. $\square$

Note, by the way, that the family of resolvents $\{J_r(z)\}_{r \geq 0}$ at the point $z \in \Delta$ for $f = I - F$ is, in fact, the rescaling approximating curve of F at this point.

Thus the question whether the mapping $I - F$ belongs to $\mathcal{G}N_\rho(\Delta)$ whenever $F \in N_\rho(\Delta)$ has been answered in the affirmative. In its turn, this fact implies another useful resolvent identity formula for the general case when $f \in \mathcal{G}N_\rho(\Delta)$.

Lemma 3.3.5 ([116]) *Let f be a continuous function in Δ which satisfies the range condition (RC), i.e., for each $s \geq 0$ the resolvent $J_s = (I + sf)^{-1}$ is well defined on Δ and belongs to $N_\rho(\Delta)$. Then for each pair $s, t \geq 0$ the mapping $G_t := (I + t(I - J_s))^{-1}$ is also well defined on Δ and belongs to $N_\rho(\Delta)$ and the following identity holds:*

$$J_{(t+1)s} = J_s \circ G_t = J_s \circ (I + t(I - J_s))^{-1} \tag{3.3.28}$$

Proof. The existence of the resolvent $G_t := (I + t(I - J_s))^{-1}$ follows directly by Proposition 3.3.1 and its Corollary 3.3.1. The same assertions imply that $G_t \in N_\rho(\Delta)$. By definition this mapping satisfies the identity:

$$(I + t(I - J_s))(G_t(z)) = z, \quad z \in \Delta. \tag{3.3.29}$$

Reminding that $I - J_s = sf(J_s)$ we obtain

$$G_t(z) + stf(J_s(G_t(z))) = z, \; z \in \Delta \tag{3.3.30}$$

for all $z \in \Delta$.

At the same time rewriting (3.3.29) in the form:

$$G_t(z) = \frac{t}{1+t} J_s(G_t(z)) + \frac{1}{1+t} z, \quad z \in \Delta,$$

we have by (3.3.30) the following identity

$$J_s\left(G_t(z)\right) + (1+t)sf\left(J_s\left(G_t(z)\right)\right) = z \qquad (3.3.31)$$

for all $z \in \Delta$. Since the equation

$$w + (1+t)\,sf\left(w\right) = z, \qquad z \in \Delta,$$

has a unique solution $w = (I + (1+t)sf)^{-1}(z) := J_{(1+t)s}(z)$, provided that (3.3.31) holds, (3.3.28) results. $\square$

Here follows the basic assertion which demonstrates the resolvent method.

Proposition 3.3.2 *Let f be a continuous function in Δ which satisfies the range condition (RC) and let $J_r = (I + rf)^{-1} \in N_\rho(\Delta)$, $r \geq 0$, be its resolvent. Then:*

(i) for each $r > 0$ the sets $\mathrm{Fix}(J_r)$ and $Null(f)$ in Δ coincide;

(ii) for each $z \in \Delta$ the net $\{J_r(z)\}_{r \geq 0}$ is convergent as $r \to \infty$.

Moreover,

(a) if $W = \mathrm{Null}(f)$ in Δ is not empty, then the limit mapping $F = \lim\limits_{r \to \infty} J_r$ is a ρ-nonexpansive retraction on W;

(b) if $W = \mathrm{Null}(f)$ in Δ is empty, then the limit mapping $F = \lim_{r \to \infty} J_r$ is a unimodular constant ζ which is the sink point for each J_r, $r > 0$. In other words, for each $K > 0$ the horocycle

$$D(\zeta, K) = \left\{ z \in \Delta : \varphi_\zeta(z) := \frac{|1 - z\bar\zeta|^2}{1 - |z|^2} < K \right\}$$

internally tangent to $\partial \Delta$ at ζ is J_r-invariant, i.e.,

$$J_r\left(D\left(\zeta, K\right)\right) \subseteq D\left(\zeta, K\right).$$

Proof. Assertion (i) follows directly by the definition of the resolvent and the uniqueness of the solution of equation (3.3.11). Thus for each $r > 0$ we have $W := \mathrm{Null}(f) = \mathrm{Fix}(J_r)$. Consider now the mapping $G_t := \left(I + s(I - J_s)^{-1}\right)$ that was introduced in the previous lemma. By the construction of this mapping we have, in turn, that for each pair $s, t > 0$ $W = \mathrm{Fix}(J_s) = \mathrm{Fix}(G_t)$. Moreover, it follows by Propositions 2.4.3 and 2.4.4 that for each $z \in \Delta$ there exists the limit $F(z) = \lim\limits_{t \to \infty} G_t(z) \in \overline\Delta$. More precisely, if $W = \mathrm{Null}(f)$ in Δ is not empty, then by Proposition 2.4.4 F is a ρ-nonexpansive retraction on W. Otherwise, F is a unimodular constant ζ which is the sink point for J_s. But in both cases equation (3.3.29) implies:

$$J_s\left(G_t(z)\right) - G_t(z) \to 0 \text{ as } t \to \infty$$

for all $z \in \Delta$. Setting now $r = (1+t)s$ and letting t to ∞ we obtain by Lemma 3.3.5 that $F = \lim\limits_{r \to \infty} J_r$. Finally, condition (b) follows from the resolvent identity (Lemma 3.3.4) and we are done. $\square$

3.4 Monotonicity with respect to the hyperbolic metric

Our main goal in the following considerations is to find analytical characterizations and parametric representations of the classes $\mathcal{G}N_\rho(\Delta)$ and $\mathcal{G}\operatorname{Hol}(\Delta)$, respectively. It turns out, that a good tool in these investigations is the notion of monotonicity with respect to the hyperbolic Poincaré metric on the unit disk.

To motivate the definition below, we recall that a mapping $f : \mathbb{R}^2 \mapsto \mathbb{R}^2$ is said to be monotone with respect to the Euclidean norm of $\mathbb{R}^2$ if for each x, y in a domain of definition f

$$(x - y, f(x) - f(y)) \geq 0, \tag{3.4.1}$$

where by $(\cdot, \cdot)$ we denote the inner scalar product in $\mathbb{R}^2$. Since each complex-valued function $f : \mathbb{C} \mapsto \mathbb{C}$ can be considered as a mapping from $\mathbb{R}^2$ into itself, condition (3.4.1) can be rewritten in the form:

$$\operatorname{Re}\left[(f(z) - f(w))\overline{(z - w)}\right] \geq 0 \tag{3.4.2}$$

for each pair z, w in a domain of definition of f. At the same time it is easy to see that (3.4.2) is equivalent to the following condition:

$$|z + rf(z) - (w + rf(w))| \geq |z - w| \tag{3.4.3}$$

for all $r \geq 0$.

The latter condition is a key to define the notion of monotonicity with respect to the Poincaré hyperbolic metric on Δ (see [116]).

Definition 3.4.1 *Let f be a complex valued function on Δ, and let ρ be the Poincaré metric on Δ. The function f is called ρ-monotone (monotone with respect to the metric ρ) if for each pair $z, w \in \Delta$ the following condition holds:*

$$\rho(z + rf(z), w + rf(w)) \geq \rho(z, w) \tag{3.4.4}$$

for all $r \geq 0$ such that $z + rf(z)$ and $w + rf(w)$ belong to Δ.

Proposition 3.4.1 *Let $f : \Delta \mapsto \mathbb{C}$ be a continuous function in Δ. Then f is ρ-monotone if and only if it satisfies the range condition (RC).*

Proof. Let f satisfy the range condition (RC), i.e., for all $r \geq 0$ the nonlinear resolvent $J_r = (I + rf)^{-1}$ is a well defined ρ-nonexpansive self-mapping of Δ:

$$\rho\left((I + rf)^{-1}(u), (I + rf)^{-1}(v)\right) \leq \rho(u, v) \tag{3.4.5}$$

for all $u, v \in \Delta$. Take now any pair z and w in Δ and let $r \geq 0$ be such that $z + rf(z) := u$ and $w + rf(w) := v$ belong to Δ. Then by definition

$$z = (I + rf)^{-1}(u) \quad \text{and} \quad w = (I + rf)^{-1}(v).$$

It is clear now that (3.4.5) implies (3.4.4), i.e., f is ρ-monotone.

Conversely. For a pair $z, w \in \Delta$ denote $u = z + f(z)$ and $v = w + f(w)$. Then for $r \geq 0$ sufficiently small (3.4.4) implies:

$$\rho\left((1 - r)z + ru, (1 - r)w + rv\right) \geq \rho(z, w). \tag{3.4.6}$$

Denoting the left hand of (3.4.6) by $\phi(r)$ we can rewrite it as

$$\phi(r) \geq \phi(0).$$

Now it follows by Proposition 2.3.2 that the latter inequality is equivalent to the condition:

$$\mathrm{Re}\left[\frac{(z - u)\bar{z}}{1 - |z|^2} + \frac{(w - v)\bar{w}}{1 - |w|^2}\right] \geq \mathrm{Re}\,\frac{\bar{z}(w - v) + w\overline{(z - u)}}{1 - \bar{z}w}.$$

Substituting $u = f(z) + z$ and $v = f(w) + w$ into this inequality we obtain a characterization of a ρ-monotone function:

$$\mathrm{Re}\left[\frac{f(z)\bar{z}}{1 - |z|^2} + \frac{f(w)\bar{w}}{1 - |w|^2}\right] \geq \mathrm{Re}\,\frac{\bar{z}f(w) + w\overline{f(z)}}{1 - \bar{z}w}. \tag{3.4.7}$$

Substituting now $z = 0$ into (3.4.7) we obtain the condition:

$$\mathrm{Re}\,f(w)\bar{w} \geq \mathrm{Re}\,f(0)\bar{w}(1 - |w|^2). \tag{3.4.8}$$

for all $w \in \Delta$.

Now we will show that condition (3.4.8) implies the solvability of the equation:

$$w + rf(w) = z \tag{3.4.9}$$

for each $r \geq 0$ and $z \in \Delta$.

To this end we will establish a more general assertion which we will call the numerical range lower bound (cf., [60] and [62]).

Lemma 3.4.1 *Let* $\alpha : [0, 1] \mapsto \mathbb{R}$ *be a continuous function on* $[0, 1]$ *such that* $\alpha(0) \leq 0$ *and the equation*

$$s + r\alpha(s) = t$$

has a unique solution $s = s(t) \in [0, 1)$ *for each* $t \in [0, 1)$ *and* $r \geq 0$. *Suppose that* $f : \Delta \mapsto \mathbb{C}$ *be a continuous function on* Δ *which satisfies the following condition:*

$$\mathrm{Re}\,f(w)\bar{w} \geq \alpha(|w|)\,|w|, \quad w \in \Delta. \tag{3.4.10}$$

Then for each $z \in \Delta$ *and* $r \geq 0$ *equation (9) has a unique solution* $w = w(z)$ *such that*

$$|w(z)| \leq s(t), \tag{3.4.11}$$

whenever $|z| \leq t < 1$.

Proof. Fix $t \in (0,1)$ and $z \in \Delta$, $|z| \leq t < 1$, and consider the equation:

$$s + r\alpha(s) = |z|.$$

It follows from our assumption that this equation has a unique solution $s_0 = s_0(|z|)$. Then setting $\gamma(s) = s + r\alpha(s) - |z|$ we have $\gamma(0) < 0$ and $\gamma(s_0) = 0$. Note that γ must be monotone on $[0,1]$. Hence, for an arbitrary $0 < \delta < 1 - s_0$ we can find $\varepsilon > 0$ such that $\gamma(s_0 + \delta) \geq \varepsilon$. Taking now $w \in \Delta$ such that $|w| = s_0 + \delta$ we have by (3.4.10) for those w:

$$\begin{aligned}
\mathrm{Re}(w + rf(w) - z)\bar{w} &\geq |w|^2 + r\alpha(|w|)|w| - |w||z| \\
&= |w|\gamma(|w|) \geq |w|\varepsilon > 0.
\end{aligned}$$

Then it follows by the Bohl–Poincaré theorem (see, for example, [79]), that equation (3.4.9) has a unique solution $w = w(z)$ such that $|w| < s_0 + \delta$. Since δ is an arbitrary sufficiently small number, we must have $|w| \leq s_0 \leq s(t)$. □

To complete the proof of Proposition 3.4.1 we first note that the function

$$\alpha(s) = -|f(0)|\left(1 - s^2\right)$$

satisfies the conditions of the above Lemma. Therefore inequality (3.4.8) and this Lemma imply that for each $r \geq 0$ the resolvent J_r, defined by $J_r(z) = w(z)$ — the solution of (3.4.9) is a single-valued self-mapping of Δ. It remains to show that this mapping is ρ-nonexpansive on Δ. Indeed, for a pair u and v in Δ, setting $z = J_r(u)$ and $w = J_r(v)$ we obtain $z + rf(z) = u$ and $w + rf(w) = v$. Since f is ρ-monotone we obtain finally

$$\rho(u,v) = \rho\left(z + rf(z), w + rf(w)\right) \geq \rho(z,w) = \rho\left(J_r(u), J_r(v)\right).$$

The Lemma is proved. □

We already mentioned in the proof of Proposition 3.4.1 that, in fact, condition (3.4.7) is equivalent to the property of a continuous function f on Δ to be ρ-monotone. Thus combining these assertions with Proposition 3.3.1 and Lemma 3.3.1 we can formulate a summary assertion for this chapter which characterizes the property of a continuous function on Δ to be in class $\mathcal{GN}_\rho(\Delta)$.

Proposition 3.4.2 ([116]) *Let f be a complex-valued continuous function on Δ. Then the following conditions are equivalent:*

(i) $f \in \mathcal{GN}_\rho(\Delta)$, i.e., it is a generator of a continuous flow $S = \{F_t\}_{t \geq 0}$ of ρ-nonexpansive self-mappings of Δ;

(ii) there is a family $\{G_t\}$ $0 \leq t \leq T$, of ρ-nonexpansive self-mappings of Δ such that

$$f(z) = \lim_{t \to 0+} \frac{z - G_t(z)}{t}$$

for each $z \in \Delta$;

(iii) the Cauchy problem

$$
\begin{cases}
\dfrac{\partial u(t,z)}{\partial t} + f\left(u(t,z)\right) = 0, \\[2ex]
u(0,z) = z \in \Delta
\end{cases}
$$

has a unique solution $u(t,z) \in \Delta$ for all $t \geq 0$ and $z \in \Delta$, such that for each $t \geq 0$ $\;u(t,\cdot) \in N_\rho(\Delta)$;

(iv) f satisfies the range condition (RC), i.e., for each $r \geq 0$ and $z \in \Delta$ the equation

$$
w + zf(w) = z
$$

has a unique solution $w = J_r(z)$, such that $J_r \in N_\rho(\Delta)$;

(v) f is a ρ-monotone function on Δ;

(vi) f satisfies the condition:

$$
\mathrm{Re}\left[\frac{f(z)\bar{z}}{1-|z|^2} + \frac{f(w)\bar{w}}{1-|w|^2}\right] \geq \mathrm{Re}\,\frac{\bar{z}f(w) + w\overline{f(z)}}{1-\bar{z}w}
$$

for each pair z, w in Δ;

Remark 3.4.1 Condition (vi) plays a crucial role in our further considerations. Inequalities of such a type (which characterize the classes of generators of flows) are called *flow invariance conditions*. For the class of holomorphic mappings a simpler inequality (3.4.8) is also a flow invariance condition, since it characterizes the class of holomorphic generators. Indeed, as we saw in the proof of Proposition 3.4.1 (or in Lemma 3.4.1) this condition is sufficient for the existence of the (nonlinear) resolvent $(I+rf)^{-1}$ which maps Δ into itself. In addition, it follows by the Implicit Function Theorem that this mapping is holomorphic (hence, ρ-nonexpansive) in Δ, so f belongs to $\mathcal{G}\,\mathrm{Hol}(\Delta)$. The necessity of this condition follows directly from condition (vi). Thus for $f \in \mathrm{Hol}(\Delta,\mathbb{C})$ inequalities (3.4.7) and (3.4.8) are equivalent.

We will see below that these conditions can be considered as forms of the Schwarz–Pick inequalities for the classes $\mathcal{G}N_\rho(\Delta)$ and $\mathcal{G}\,\mathrm{Hol}(\Delta)$ of generators of flows of ρ-nonexpansive and holomorphic mappings, respectively.

A geometric nature of these conditions for holomorphic functions will be explained in the next section.

Here we note that an immediate consequence of these flow invariance conditions is the following (cf., Remark 3.3.1):

Corollary 3.4.1 *The sets $\mathcal{G}N_\rho(\Delta)$ and $\mathcal{G}\,\mathrm{Hol}(\Delta)$ are closed (with respect to the topology of uniform convergence on compact subsets of Δ) real cones.*

Exercise 1. Show that the set $\mathcal{G}\,\mathrm{Hol}(\Delta) \cap (-\mathcal{G}\,\mathrm{Hol}(\Delta))$ is precisely the set of all generators of one-parameter groups of automorphisms of Δ. Hence, this set is a *real vector space*.

Exercise 2. Describe the set $\mathcal{G}N_\rho(\Delta) \cap (-\mathcal{G}N_\rho(\Delta))$.

3.5 Flow invariance conditions for holomorphic functions

In this section we study several flow invariance conditions for the class of holomorphic functions. We will use these conditions to obtain parametric representation of functions of the class $\mathcal{G}\operatorname{Hol}(\Delta)$, to study their dynamic transformations and the asymptotic behavior of the flows generated by them.

In the first step we give a simpler explanation of the necessity of the flow invariance condition (3.4.8) for a function $f \in \mathcal{G}\operatorname{Hol}(\Delta)$ and later we study it in greater detail. In addition, we will see below that this condition can be improved by a more qualified condition (see Proposition 3.5.3) which has some additional applications. Also, note that in the case of holomorphic functions one uses a special terminology which comes from the theory of bounded symmetric domains (see, for example, [142, 12, 32]).

Definition 3.5.1 *A function $f \in \operatorname{Hol}(\Delta, \mathbb{C})$ is said to be a semi-complete vector field on Δ if the Cauchy problem (*):*

$$
\begin{cases}
\dfrac{\partial u(t, z)}{\partial t} + f\left(u(t, z)\right) = 0, \\[2mm]
u(0, z) = z \in \Delta,
\end{cases}
\tag{*}
$$

has a unique solution $u(t, z) \in \Delta$ for all $z \in \Delta$ and all nonnegative t, i.e., $t \in \mathbb{R}^+ = [0, \infty)$.

If the Cauchy problem (*) has a solution $u(t, z)$ defined for all real t, i.e., $t \in \mathbb{R} = (-\infty, \infty)$, then f is said to be complete (or integrated) (see, for example, [32, 142]).

Thus f is semi-complete if and only if it is an infinitesimal generator of a one-parameter semigroup; f is complete if and only if it is a generator of a one-parameter group.

Exercise 1. Show that the mapping $f : \Delta \mapsto \mathbb{C}$, defined as $f(z) = z - z^2$ is a semi-complete vector field on Δ.

Exercise 2. Show that $f : \Delta \mapsto \mathbb{C}$, defined by $f(z) = z - 1 + \sqrt{1 - z}$ is a semi-complete vector field. Find the flow generated by f.

Exercise 3. Show that for each $a \in \mathbb{C}$ the mapping $f : f(z) = a - \bar{a}z^2$ is a complete vector field.

We begin with a characterization of all complete vector fields. Note again that if f is a complete vector field, then it generates a one-parameter subgroup $S = \{F_t\}_{t \in \mathbb{R}}$ of the group $\operatorname{Aut}(\Delta)$ of all automorphisms of Δ. Consequently, each

$F_t \in S$ is a fractional linear Möbius transformation. Thus F_t has a holomorphic continuation on a neighborhood of $\overline{\Delta}$ and so does f, because of the equality:

$$f = \lim_{t \to 0^+} \frac{I - F_t}{t}. \tag{3.5.1}$$

The family of complete vector fields on Δ will be denoted by $\text{aut}(\Delta)$. As we mentioned above, this family is a real vector space, which can be described as follows.

Since $|F_t(z)| \le 1$, for $z \in \overline{\Delta}$ we have by (3.5.1) that:

$$\text{Re} f(z)\bar{z} \ge 0 \quad \text{for all } z \in \partial\Delta. \tag{3.5.2}$$

At the same time the function $-f$ is also a complete vector field on Δ. Therefore we also obtain that

$$-\text{Re} f(z)\bar{z} \ge 0 \quad \text{for all } z \in \partial\Delta. \tag{3.5.3}$$

Comparing (3.5.2) and (3.5.3) we obtain the necessary boundary condition for $f \in \text{aut}(\Delta)$:

$$\text{Re} f(z)\bar{z} = 0 \quad \text{for all } \quad z \in \partial\Delta \tag{3.5.4}$$

(see Figure 3.1)

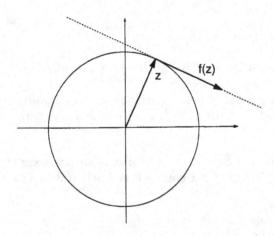

Figure 3.1: Boundary condition for $f \in \text{aut}(\Delta)$.

Actually, this condition is also sufficient for $f \in \text{Hol}(\overline{\Delta}, \mathbb{C})$ to be complete. Indeed, suppose that $f \in \text{Hol}(\overline{\Delta}, \mathbb{C})$ and satisfies (3.5.4). Rewriting f in the Taylor series form:

$$f(z) = a_0 + a_1 z + a_2 z^2 + \dots$$

we have for $z \in \partial\Delta$:

$$\text{Re} f(z)\bar{z} = \text{Re} g(z) = 0,$$

where
$$g(z) = a_1 + (\overline{a_0} + a_2)z + a_3 z^2 + \ldots \in \mathrm{Hol}(\overline{\Delta}, \mathbb{C}).$$

It follows by the maximum principle for harmonic functions that:
$$\mathrm{Re}\, a_1 = \overline{a_0} + a_2 = a_3 = \ldots = 0.$$

Hence $f(z)$ is actually polynomial of the second order at most:
$$f(z) = a_0 + a_1 z + a_2 z^2 \qquad (3.5.5)$$

with
$$\overline{a_0} = -a_2, \quad \text{and} \quad \mathrm{Re}\, a_1 = 0. \qquad (3.5.6)$$

Now it can be shown by direct computation that the Cauchy problem:
$$\begin{cases} \dfrac{\partial u(t,z)}{\partial t} + f\left(u(t,z)\right) = 0, \\ u(0,z) = z \in \Delta, \end{cases}$$

with f satisfying (3.5.5) and (3.5.6) has a unique solution $u(t, \cdot) \in \mathrm{Hol}(\Delta)$ for all $t \in \mathbb{R}$. Moreover, for fixed $t \in \mathbb{R}$ the mapping $F_t = u(t, \cdot)$ is a Möbius transformation of the unit disk.

So, we have proved the following result (see, for example, [16, 8]).

Proposition 3.5.1 (Boundary group invariance condition) *A mapping $f \in \mathrm{Hol}(\Delta, \mathbb{C})$ is a complete vector field on Δ (i.e., $f \in \mathrm{aut}(\Delta)$) if and only if it has a continuous extension to $\overline{\Delta}$ and*

$$\mathrm{Re}\, f(z)\bar{z} = 0 \quad \text{for all } z \in \partial\Delta.$$

This is equivalent to the statement that f is a polynomial of the second order at most:
$$f(z) = a_0 + a_1 z + a_2 z^2$$
with coefficients a_0, a_1, a_2 which satisfy the conditions:

$$\mathrm{Re}\, a_1 = 0, \quad \bar{a}_0 = -a_2.$$

Corollary 3.5.1 *The family $\mathrm{aut}(\Delta)$ is a real vector space of entire functions. Moreover, this space has the following decomposition:*

$$\mathrm{aut}(\Delta) = \mathrm{aut}_0(\Delta) \oplus P_2,$$

where

$$\mathrm{aut}_0(\Delta) = \{f \in \mathrm{aut}(\Delta) : f(0) = 0\}$$
$$= \{f \in \mathrm{Hol}(\Delta, \mathbb{C}) : f(z) = bz, \ \mathrm{Re}\, b = 0\}$$

is the subspace of linear functions, and

$$P_2 = \{f \in \mathrm{aut}(\Delta), \ f'(0) = 0\}$$
$$= \{f \in \mathrm{Hol}(\Delta, \mathbb{C}) : f(z) = \bar{a} - az^2, \ a \in \mathbb{C}\}$$

is the subspace of so called 'transvections'.

Now we will turn to the general case when $f \in \mathrm{Hol}(\Delta, \mathbb{C})$ is a semi-complete vector field on the unit disk. To explain the nature of the condition (3.4.8) we adduce first some heuristic grasps. Assume temporarily that $f \in \mathcal{G}\,\mathrm{Hol}(\Delta)$ is holomorphic in the neighborhood of Δ. In this case we will write just $f \in \mathrm{Hol}(\overline{\Delta}, \mathbb{C})$. Then, again we have the following boundary flow invariance condition:

$$\mathrm{Re}\, f(z) \cdot \bar{z} \geq 0 \quad \text{for all} \quad z \in \partial\Delta$$

(see Figure 3.2).

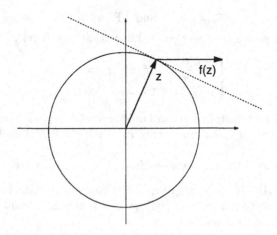

Figure 3.2: Boundary flow invariance condition.

It implies that

$$\mathrm{Re}(f(z) - f(0))\bar{z} \geq -\,\mathrm{Re}\, f(0)\bar{z}, \quad z \in \partial\Delta.$$

Dividing the left hand side of this inequality by $|z|^2 = 1$ we obtain:

$$\mathrm{Re}\left(\frac{f(z) - f(0)}{z}\right) \geq -\,\mathrm{Re}\,\overline{f(0)}z, \quad z \in \partial\Delta.$$

Now again it follows by the maximum principle for harmonic functions that the latter inequality holds also for $z \in \Delta$. Multiplying it by $|z|^2 \neq 0$, $z \in \Delta$, we obtain:

$$\mathrm{Re}\, f(z) \cdot \bar{z} \geq \mathrm{Re}\, f(0) \cdot \bar{z} \cdot (1 - |z|^2), \quad z \in \Delta. \tag{3.5.7}$$

However, there are holomorphic mappings on the unit disk which are semi-complete, but have no holomorphic extension to $\overline{\Delta}$, the closure of Δ. Consider, for example,

$$f(z) = z - 1 + \sqrt{1 - z}.$$

Nevertheless, we will see that even $f \in \mathrm{Hol}(\Delta, \mathbb{C})$ does not extend continuously to $\overline{\Delta}$, condition (3.5.7) is necessary and sufficient for f to be in $\mathcal{G}\,\mathrm{Hol}(\Delta)$. The necessity can be shown also by the following simple considerations which are useful, however, to obtain a parametric representation of the class $\mathcal{G}\,\mathrm{Hol}(\Delta)$. As we already know this class is a real cone. Therefore, if we present $f \in \mathcal{G}\,\mathrm{Hol}(\Delta)$ in the form:

$$f(z) = g(z) + h(z), \tag{3.5.8}$$

where

$$g(z) = f(0) - \overline{f(0)}z^2$$

is a transvection (i.e., $g \in P_2$, see Corollary 3.5.1), we have

$$h(z) = f(z) + (-g(z)) \in \mathcal{G}\,\mathrm{Hol}(\Delta) \tag{3.5.9}$$

and

$$h(0) = 0. \tag{3.5.10}$$

Now, conditions (3.5.9) and (3.5.10) imply that there is a semigroup $S_h = \{H_t\}_{t \geq 0}$ of holomorphic self-mappings H_t of Δ, such that $H_t(0) = 0$, for all $t \geq 0$. Hence by the Schwarz Lemma we have:

$$|H_t(z)| \leq |z|, \quad \text{for all } z \in \Delta.$$

Since

$$h(z) = \lim_{t \to 0^+} \frac{1}{t}(z - H_t(z))$$

we obtain:

$$\mathrm{Re}\, h(z)\bar{z} \geq 0 \quad \text{for all } z \in \Delta. \tag{3.5.11}$$

In addition, note that

$$\mathrm{Re}\, g(z)\bar{z} = \mathrm{Re}\, f(0)\bar{z} \cdot (1 - |z|^2). \tag{3.5.12}$$

Then by (3.5.11), (3.5.12), and (3.5.8) we obtain (3.5.7) and the necessity of this condition is proved.

The sufficiency of condition (3.5.7) for $f \in \mathrm{Hol}(\Delta, \mathbb{C})$ to be a semi-complete vector field was established in Lemma 3.4.1 (see, also, Remark 3.4.1). However, for the case of holomorphic functions one can make this lemma more precise.

Lemma 3.5.1 *Let $\alpha \in \mathcal{G}\,\mathrm{Hol}(\Delta)$ be such that $\alpha(|z|) \in \mathbb{R}$, $z \in \Delta$, and let $f \in \mathrm{Hol}(\Delta, \mathbb{C})$ satisfy the following condition:*

$$\mathrm{Re}\, f(z)\bar{z} \geq \alpha(|z|)\,|z|, \quad z \in \Delta.$$

Then:

(i) f is a semi-complete vector field on Δ;

(ii) if $S_f = \{F_t\}$, $t \geq 0$, is the semigroup generated by f, then for all $t \geq 0$ and $x \in \Delta$

$$|F_t(x)| \leq \beta_t(|x|),$$

where β_t is the solution of the Cauchy problem:

$$\begin{cases} \dfrac{d\beta_t(s)}{dt} + \alpha(\beta_t(s)) = 0, \\ \beta_0(s) = s, \quad s \in [0,1). \end{cases}$$

Proof. It follows by Proposition 3.3.1 that the function $\alpha : \Delta \mapsto \mathbb{C}$ satisfies the range condition, i.e., the equation: $w + r\alpha(w) = z$ has a unique solution $w = (I + r\alpha)^{-1}(z)$ for each $r \geq 0$ and $z \in \Delta$. Since on the interval $[0,1)$ the function α is real-valued, it is not difficult to see that $s = (I + r\alpha)^{-1}(t) \in [0,1)$ whenever $t \in [0,1)$. It then follows by Lemma 3.4.1 that for a fixed $r \in (0,1)$ and $z \in \Delta$ the equation: $w + rf(w) = z$ has a unique solution $w = (I + rf)^{-1}(z)$ and

$$\left|(I + rf)^{-1}(z)\right| \leq (I + r\alpha)^{-1}(|z|).$$

Using again Proposition 3.3.1 and the exponential formula we obtain our assertion. $\square$

Now, observing that the function:

$$\alpha(z) = -|f(0)|(1 - z^2)$$

satisfies all the conditions of Lemma 3.5.1 we obtain the following result:

Proposition 3.5.2 (see [7]) *Let $f \in \mathrm{Hol}(\Delta, \mathbb{C})$. Then f is a semi-complete vector field if and only if it satisfies condition (3.5.7):*

$$\mathrm{Re}\, f(z)\bar{z} \geq \mathrm{Re}\, f(0)\bar{z}(1 - |z|^2)$$

for all $z \in \Delta$.

Moreover, if $S_f = \{F_t\}$, $t \geq 0$, is a semigroup generated by f, then the following estimate of growth holds:

$$|F_t(z)| \leq \frac{|z| + 1 - e^{-2|f(0)|t}(1 - |z|)}{|z| + 1 + e^{-2|f(0)|t}(1 - |z|)}. \tag{3.5.13}$$

Remark 3.5.1 Note that *if equality in (3.5.7) holds for at least one value $z_0 \in \Delta$ then it holds for all $z \in \Delta$, and $f(z)$ is actually a complete vector field.*

Indeed, by (3.5.8) we have:
$$f = g + h,$$

where $g \in \mathrm{aut}(\Delta)$, and $h \in \mathcal{G}\,\mathrm{Hol}(\Delta)$, with $h(0) = 0$.

Therefore if we present h in the form

$$h(z) = z \cdot p(z),$$

we obtain that $p \in \mathrm{Hol}(\Delta, \mathbb{C})$ and

$$\mathrm{Re}\, p(z) \geq 0, \quad z \in \Delta. \tag{3.5.14}$$

Thus (3.5.7) is equivalent to the following equation

$$f(z) = f(0) + zp(z) - \overline{f(0)}z^2.$$

If now for some $z_0 \in \Delta$ we have equality in (3.5.11) we also have the equality $\mathrm{Re}\, p(z_0) = 0$. It then follows by the maximum principle that $p(z) = p = \mathrm{const}$, hence $f(z)$ has the form (3.4.5)

$$f(z) = f(0) + ipz - \overline{f(0)}z^2,$$

i.e., it is a complete vector field.

As a matter of fact (3.5.14) enables us to find a more qualified estimate which also characterizes semi-complete vector fields. To do this we need the following assertion.

Lemma 3.5.2 (Harnack inequality) *If $p \in \mathrm{Hol}(\Delta, \mathbb{C})$ maps Δ into the right half-plane, i.e.,*

$$\mathrm{Re}\, p(z) \geq 0, \quad z \in \Delta,$$

then it satisfies the strong estimate:

$$\mathrm{Re}\, p(0)\frac{1 - |z|}{1 + |z|} \leq \mathrm{Re}\, p(z) \leq \mathrm{Re}\, p(0)\frac{1 + |z|}{1 - |z|}. \tag{3.5.15}$$

Proof. If $\mathrm{Re}\, p(0) = 0$, then as above $\mathrm{Re}\, p(z) = 0$ for all $z \in \Delta$ and (3.5.15) is obvious. If $p \in \mathrm{Hol}(\Delta, \mathbb{C})$ satisfies the strong inequality in (3.5.14), then the function p_1:

$$p_1(z) = \frac{1}{\mathrm{Re}\, p(0)}[p(z) - i\,\mathrm{Im}\, p(0)]$$

belongs to the so called class of Carathéodory:

$$\mathrm{Re}\, p_1(z) > 0, \; z \in \Delta \quad \text{and} \quad p_1(0) = 1.$$

Since the fractional linear transformation:

$$G(w) = \frac{w - 1}{w + 1} \tag{3.5.16}$$

maps the right half-plane $\{w \in \mathbb{C} : \mathrm{Re}\, w > 0\}$ into the unit disk (see Exercise 4 bellow), we have that the mapping F:

$$F(z) = G(p_1(z))$$

is a holomorphic self-mapping of Δ and $F(0) = 0$. Then the Schwarz Lemma implies that for each $0 \leq r < 1$ and $z \in \Delta : |z| \leq r$ the following inequality holds:

$$\left|\frac{p_1(z) - 1}{p_1(z) + 1}\right| \leq r.$$

Now it is easy to see that the circle $\left|\dfrac{w - 1}{w + 1}\right| = r$ is symmetric with respect to the real axis and intersects it at the points $\dfrac{1 + r}{1 - r}$ and $\dfrac{1 - r}{1 + r}$ (see Figure 3.3).

Therefore we obtain the following inequality for each function p_1 of the class of Carathéodory:

$$\frac{1 - |z|}{1 + |z|} \leq \mathrm{Re}\, p_1(z) \leq \frac{1 + |z|}{1 - |z|}.$$

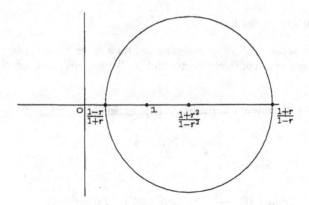

Figure 3.3: Values of functions of Carathéodory's class.

Since
$$\operatorname{Re} p_1(z) = \frac{1}{\operatorname{Re} p(0)} \operatorname{Re} p(z)$$

we obtain (3.5.15). □

Exercise 4. Prove that for all w such that $\operatorname{Re} w \geq 0$ ($w \in \Pi_+$) the following inequality holds:
$$\left| \frac{w-1}{w+1} \right| \leq 1.$$

Exercise 5. Prove the equivalence of the following assertions:

(a) a function $p \in \operatorname{Hol}(\Delta, \mathbb{C})$ has values (for all $z \in \Delta$) in a compact subset of the open right-half plane $\{w \in \mathbb{C} : \operatorname{Re} w > 0\}$;

(b) $p(z) = \dfrac{1 - \omega(z)}{1 + \omega(z)}$, where $\omega \in \operatorname{Hol}(\Delta)$ is such that $|\omega(z)| \leq c < 1$ for all $z \in \Delta$.

Exercise 6. Show that under the conditions of Exercise 5 the function p satisfies the **strong Harnack inequality**:
$$\operatorname{Re} p(0) \frac{1 + c|z|}{1 - c|z|} \geq \operatorname{Re} p(z) \geq \operatorname{Re} p(0) \frac{1 - c|z|}{1 + c|z|}, \quad 0 \leq c < 1.$$

Finally, observe that from the representation of $f \in \mathcal{G} \operatorname{Hol}(\Delta)$:
$$f(z) = f(0) - \overline{f(0)}z^2 + z \cdot p(z), \quad \operatorname{Re} p(z) \geq 0 \qquad (3.5.17)$$

we deduce that:
$$p(0) = f'(0).$$

Thus we have proved the following assertion.

Proposition 3.5.3 ([6]) *A function $f \in \mathrm{Hol}(\Delta, \mathbb{C})$ is a semi-complete vector field on Δ if and only if $\mathrm{Re}\, f'(0) \geq 0$ and the following inequality holds:*

$$\mathrm{Re}\, f'(0) \frac{|z|^2(1+|z|)}{1-|z|} + \mathrm{Re}\, f(0)\bar{z}(1-|z|^2) \geq \mathrm{Re}\, f(z)\bar{z}$$

$$\geq \mathrm{Re}\, f(0)\bar{z}(1-|z|^2) + \mathrm{Re}\, f'(0) \frac{|z|^2(1-|z|)}{1+|z|}. \qquad (3.5.18)$$

Moreover, the equality in (3.5.16) holds if and only if $\mathrm{Re}\, f'(0) = 0$.

Corollary 3.5.2 ([7]) *A mapping $f \in \mathcal{G}\,\mathrm{Hol}(\Delta)$ belongs to $\mathrm{aut}(\Delta)$ if and only if*

$$\mathrm{Re}\, f'(0) = 0.$$

Corollary 3.5.3 ([7]) *If $f \in \mathcal{G}\,\mathrm{Hol}(\Delta)$ is given with $f(0) = f'(0) = 0$, then $f \equiv 0$.*

This assertion can be considered as a tangential version of the Schwarz Lemma. Indeed, if $F \in \mathrm{Hol}(\Delta)$ is a holomorphic self-mapping, then $f = I - F$ is a semi-complete vector field. Therefore, if $F(0) = 0$ and $F'(0) = 1$, then $f(0) = f'(0) = 0$. Therefore, by the Corollary 3.4.2 we obtain that $F(z) \equiv z$. This is the statement of the second part of the Schwarz Lemma.

Recently M. Abate [5] established another condition, which characterizes a semi-complete vector field f by using the estimate for its derivative f'.

To establish his condition we first prove the following characterization of the class $\mathcal{P} := \{p \in \mathrm{Hol}(\Delta, \mathbb{C}) : \mathrm{Re}\, p(z) \geq 0, \ z \in \Delta\}$.

Lemma 3.5.3 ([7]) *Let $p \in \mathrm{Hol}(\Delta, \mathbb{C})$. Then condition (3.5.14):*

$$\mathrm{Re}\, p(z) \geq 0,$$

holds for all $z \in \Delta$, if and only if there is a positive function $\psi : [0,1) \mapsto \mathbb{R}^+$ such that :

$$\mathrm{Re}\,(zp'(z) + \psi(|z|)p(z)) \geq 0 \qquad (3.5.19)$$

for all $z \in \Delta$.

Proof. Let $p \in \mathrm{Hol}(\Delta, \mathbb{C})$ satisfy (3.5.14). Define $F = (p-1)(p+1)^{-1}$ which maps Δ into itself, $F \in \mathrm{Hol}(\Delta)$. Applying the Schwarz–Pick Lemma to F we obtain the inequality:

$$\left| \left(\frac{p-1}{p+1} \right)' \right| = \frac{2|p'|}{|1+p|^2} \leq \frac{|p+1|^2 - |p-1|^2}{|p+1|^2(1-|z|^2)},$$

which implies:

$$|p'(z)| \leq \frac{2\,\mathrm{Re}\, p(z)}{1-|z|^2}.$$

Consequently,

$$\mathrm{Re}(-zp'(z)) \le |zp'(z)| \le \frac{2|z|\,\mathrm{Re}\,p(z)}{1-|z|^2} \le \frac{1+|z|^2}{1-|z|^2}\,\mathrm{Re}\,p(z).$$

Setting here $\psi(t) = \dfrac{1+t^2}{1-t^2}$ we obtain (3.5.19).

Conversely. Suppose that (3.5.19) holds with a positive $\psi : [0,1) \mapsto \mathbb{R}^+$.
Setting $z = re^{i\theta}$ we have:

$$zp'(z) = r\frac{\partial p}{\partial r}$$

and (3.5.19) becomes:

$$\left(\mathrm{Re}\left(r\frac{\partial p}{\partial r}\right) + \psi(r)\,\mathrm{Re}\,p(z)\right) \ge 0, \quad z = re^{i\theta} \in \Delta. \tag{3.5.20}$$

Assume that there exists $z_0 = r_0 e^{i\theta_0}$ in Δ such that $\mathrm{Re}\,p(z_0) < 0$.

But (3.5.20) implies that $\mathrm{Re}\,p(0) \ge 0$, hence there is $r_1 \in [0, r_0)$ such that $\mathrm{Re}\,p(r_1 e^{i\theta_0}) = 0$ due to continuity.

Then one can find $r_2 \in (r_1, r_0)$ such that

$$\mathrm{Re}\,p(r_2 e^{i\theta_0}) < 0$$

and

$$\mathrm{Re}\,\frac{\partial p}{\partial r}(r_2 e^{i\theta_0}) < 0.$$

But these inequalities contradict (3.5.20). Thus it follows that $\mathrm{Re}\,p(z) \ge 0$ everywhere and we are done. $\square$

Now it is easy to verify that condition (3.5.19) with $\psi(r) = \dfrac{1+r^2}{1-r^2}$ is equivalent to the condition:

$$\mathrm{Re}\left[2f(z)\bar{z} + f'(z)(1-|z|^2)\right] \ge 0.$$

Thus we can summarize the assertions of this section in the following result.

Proposition 3.5.4 *Let* $f \in \mathrm{Hol}(\Delta, \mathbb{C})$. *Then the following are equivalent:*

(i) $f \in \mathcal{G}\,\mathrm{Hol}(\Delta, \mathbb{C})$, *i.e.,* f *is a semi-complete vector field on* Δ;
(ii) $\mathrm{Re}\,f(z)\bar{z} \ge \mathrm{Re}\,f(0)\bar{z}(1-|z|^2)$;
(iii) $\mathrm{Re}\,f'(0) \ge 0$ *and*

$$\mathrm{Re}\left[f(0)\bar{z}(1 - |z|^2) + f'(0)|z|^2\frac{1+|z|}{1-|z|}\right]$$
$$\ge \mathrm{Re}\,f(z)\bar{z}$$
$$\ge \mathrm{Re}\left[f(0)\bar{z}(1-|z|^2) + f'(0)|z|^2\frac{1-|z|}{1+|z|}\right];$$

(iv) $\mathrm{Re}\left[2f(z)\bar{z} + f'(z)(1-|z|^2)\right] \ge 0$;

(v) $f(z) = f(0) - \overline{f(0)}z^2 + z \cdot p(z)$, *with* $p \in \text{Hol}(\Delta, \mathbb{C})$, $\text{Re}\, p(z) \geq 0$.

Moreover, if for some $z_0 \in \Delta$ *the equality in one of the conditions (ii) or (iv) holds, then it holds for all of these conditions and for all* $z \in \Delta$. *In this case the function* $p(z)$ *in (v) is constant and* f *is actually a complete vector field.*

Remark 3.5.3 The class of functions of the form: $h(z) = z \cdot p(z)$, where $\text{Re}\, p(z) \geq 0$, $z \in \Delta$ usually referred to as class $\mathcal{N}$. The class $\mathcal{M}$ consists of all elements of $\mathcal{N}$ which are not linear functions, i.e.,

$$\mathcal{M} = \{ h \in \text{Hol}(\Delta, \mathbb{C}) : \ h(z) = z \cdot p(z), \ \text{Re}\, p(z) > 0, \ z \in \Delta \}.$$

Note that $\mathcal{N} = \text{aut}_0(\Delta) \oplus \mathcal{M}$. Thus condition (v) of the above proposition means that the *class of the semi-complete vector fields on* Δ admits the following *decompositions:*

$$\mathcal{G}\,\text{Hol}(\Delta) = P_2 + \mathcal{N} = \text{aut}(\Delta) \oplus \mathcal{M}. \tag{3.5.21}$$

(see Corollary 3.5.1).

In addition, it is well known (see, for example, [57] and [122]) that for each holomorphic function p on Δ with values in the closed right half-plane Π_+ (i.e., $\text{Re}\, p(z) \geq 0$) there exists a positive increasing finite function μ_p on the unit circle $\partial \Delta$, such that

$$p(z) = \int\limits_{\partial\Delta} \frac{1 + z\bar{\zeta}}{1 - z\bar{\zeta}} d\mu_p(\zeta) + ib \tag{3.5.22}$$

with some real b.

This formula is called the Riesz–Herglotz representation of functions in $\mathcal{P} = \{ p \in \text{Hol}(\Delta, \mathbb{C}) : \text{Re}\, p(z) \geq 0 \}$. It establishes a linear one-to-one correspondence between the set of all positive measures on $\partial \Delta$ and $\mathcal{P}$.

We will call the function $\mu_p : \partial \Delta \mapsto \mathbb{R}$ the measure characteristic function for $p \in \mathcal{P}$.

Thus by (3.5.21) and (3.5.22) we have the following *integral parametric representation for* $f \in \mathcal{G}\,\text{Hol}(\Delta)$:

$$f(z) = a - \bar{a}z^2 + izb + \int\limits_{\partial\Delta} z\frac{1 + z\bar{\zeta}}{1 - z\bar{\zeta}} d\mu(\zeta), \tag{3.5.23}$$

where $a \in \mathbb{C}$, $b \in \mathbb{R}$ and μ is a positive function on $\partial \Delta$.

Another parametric representation of the class $\mathcal{G}\,\text{Hol}(\Delta)$ which is determined by the location of null points of $f \in \mathcal{G}\,\text{Hol}(\Delta)$ is due to E. Berkson and H. Porta [17]. We will give it in the next chapter.

Exercise 7. Let $h \in \text{Hol}(\Delta, \mathbb{C})$ belong to class $\mathcal{N} : h(z) = z \cdot p(z)$, $\text{Re}\, p(z) \geq 0$, $z \in \Delta$.

(a) Show directly that the Cauchy problem:

$$\begin{cases} \dfrac{\partial u(t, z)}{\partial t} + h\left(u(t, z)\right) = 0, \\ u(0, z) = z \in \Delta, \end{cases} \tag{3.5.24}$$

has a unique solution $u(t,z) \in \Delta$ for all $t \geq 0$, and $u(t,0) = 0$, $t \geq 0$.

(b) Show that
$$\lim_{t \to \infty} u(t,z) = 0$$

if and only if $h \in \mathcal{M}$, i.e., $\operatorname{Re} p(z) > 0$, $z \in \Delta$.

(c) Show that if $h \in M$ with $h'(0) = 1$, then the limit

$$\lim_{t \to \infty} e^t u(t,z) := F(z) \tag{3.5.25}$$

exists for all $z \in \Delta$ and $F \in \operatorname{Hol}(\Delta, \mathbb{C})$.

Hint: Define the function $p_1 \in M$ by $p_1(z) = 1/p(z)$ and show that the Cauchy problem (3.5.24) is equivalent to the following integral equation

$$\ln\left[e^t u(t,z)\right] = \ln z - \int_z^{u(t,z)} (p_1(\zeta) - 1)\, \frac{d\zeta}{\zeta},$$

which implies

$$\lim_{t \to \infty} e^t u(t,z) = z \exp\left(\int_0^z (p_1(\zeta) - 1)\, \frac{d\zeta}{\zeta} \right). \tag{3.5.26}$$

(d) Show that $F \in \operatorname{Hol}(\Delta, \mathbb{C})$ defined by (3.5.25) (or (3.5.26)) satisfies the following inequality:

$$\operatorname{Re}\left[\frac{zF'(z)}{F(z)} \right] > 0. \tag{3.5.27}$$

(It is well known due to R. Nevanlinna [102] that the latter inequality characterizes all univalent F functions on Δ, normalized by $F(0) = 0$, $F'(0) \neq 0$, whose image $F(\Delta)$ is starlike with respect to zero. See Chapter 5.)

Exercise 8. Show that the function f, defined by:

$$h(z) = z\frac{1 + ze^{i\theta}}{1 - ze^{i\theta}}, \qquad \theta \in [0, 2\pi]$$

belongs to $\mathcal{M}$, and $h'(0) = 1$. Find explicitly the function $F(z)$ in (3.5.25).

Exercise 9. Show that if the function f defined as: $f(z) = a_0 + a_1 z + \sum_{k=3}^{\infty} a_k z^k$ is a semi-complete vector field on Δ and $\operatorname{Re} a_1 = 0$, then $a_0 = a_k = 0$ for all $k \geq 3$.

3.6 The Berkson–Porta parametric representation of semi-complete vector fields

An important consequence of Proposition 3.4.1 and Corollary 3.4.1 is the following representation of semi-complete vector field, which is originally due to E. Berkson and H. Porta [17] (see, also [6]).

Proposition 3.6.1 *A mapping $f \in \mathrm{Hol}(\Delta, \mathbb{C})$ is a semi-complete vector field on Δ if and only if there is a point $\tau \in \overline{\Delta}$ and a function $p \in \mathrm{Hol}(\Delta, \mathbb{C})$ with $\mathrm{Re}\, p(z) \geq 0$ everywhere such that:*

$$f(z) = (z - \tau)(1 - z\bar{\tau})p(z). \tag{3.6.1}$$

Moreover, such a representation is unique and τ is either a null point of f in Δ, or the boundary sink point of the resolvent $J_r := (I + rf)^{-1}$, $r > 0$.

Proof. Firstly, let f be a semi-complete vector field on Δ with a null-point $\tau \in \Delta$. Then it follows by formula (3.4.7) that the following inequality holds:

$$\mathrm{Re}\, f(w)\bar{w} \geq (1 - |w|^2)\,\mathrm{Re}\,\frac{f(w)\bar{\tau}}{1 - w\bar{\tau}} \tag{3.6.2}$$

or,

$$\mathrm{Re}\, f(w)\left(\frac{\bar{w}}{1 - |w|^2} - \frac{\bar{\tau}}{1 - w\bar{\tau}}\right) \geq 0. \tag{3.6.3}$$

We calculate

$$
\begin{aligned}
\frac{\bar{w}}{1 - |w|^2} - \frac{\bar{\tau}}{1 - w\bar{\tau}} &= \frac{\bar{w} - |w|^2\bar{\tau} - \bar{\tau} + \bar{\tau}|w|^2}{(1 - |w|^2)(1 - w\bar{\tau})} \\
&= \frac{\bar{w} - \bar{\tau}}{(1 - |w|^2)(1 - w\bar{\tau})} \\
&= \frac{|\bar{w} - \bar{\tau}|}{(1 - |w|^2)} \cdot \frac{1}{(w - \tau)(1 - w\bar{\tau})}.
\end{aligned}
\tag{3.6.4}
$$

Now by identifying $w \in \Delta$ with $z \in \Delta$ we obtain from (3.6.3) and (3.6.4):

$$\mathrm{Re}\,\frac{f(z)}{(z - \tau)(1 - z\bar{\tau})} \geq 0 \tag{3.6.5}$$

for all $z \in \Delta$. Denoting:

$$p(z) = \frac{f(z)}{(z - \tau)(1 - z\bar{\tau})} \tag{3.6.6}$$

we have (3.6.1).

Now suppose that $f \in \mathcal{G}\,\mathrm{Hol}(\Delta)$ has no null point in Δ. Then it follows by Proposition 3.3.2 that in this case there is a unique boundary point τ, such that for each $w \in \Delta$, the net $\{z_r(w)\}_{r>0}$ defined as the solution of the equation

$$z_r(w) + rf(z_r(w)) = w \tag{3.6.7}$$

converges to τ, as $r \to \infty$. (Indeed, for each $r \geq 0$ the value $z_r(w)$ is just the value of the resolvent $J_r = (I + rf)^{-1}$ at the point $w \in \Delta$).

Fix $\varepsilon > 0$ and consider the mapping $f_\varepsilon \in \text{Hol}(\Delta, \mathbb{C})$ defined as $f_\varepsilon(z) = \varepsilon \cdot z + f(z)$. It is clear that f_ε converges to f as ε goes to zero. Since $\mathcal{G}\,\text{Hol}(\Delta)$ is a real cone, it follows that $f_\varepsilon \in \mathcal{G}\,\text{Hol}(\Delta)$ for each $\varepsilon \geq 0$. In addition, the equation

$$f_\varepsilon(z) = 0$$

is a particular case of (3.6.7) with $r = 1/\varepsilon$ and $w = 0$. Hence f_ε has a unique null point $\tau_\varepsilon \in \Delta$ and the net $\{\tau_\varepsilon\}_{\varepsilon>0}$ converges to τ as ε tends to zero.

Since f_ε satisfies the inequality:

$$\text{Re}\,\frac{f_\varepsilon(z)}{(1 - z\tau_\varepsilon)(z - \tau_\varepsilon)} \geq 0,$$

letting ε tend to zero we have the same (inequality (3.6.5)) for $f(z)$, which in turn implies representation (3.6.1).

Conversely. Suppose that $f \in \text{Hol}(\Delta)$ admits representation (3.6.1) with $\tau \in \overline{\Delta}$ and $\text{Re}\,p(z) \geq 0$ everywhere.

If $\text{Re}\,p(z) = 0$ for some $z \in \Delta$, then by the maximum principle it follows that $p(z) = im$ for some $m \in \mathbb{R}$. In this case:

$$
\begin{aligned}
f(z) &= (z - \tau)(1 - z\bar{\tau})im = (z - \tau - z^2\bar{\tau} + z|\tau|^2) \\
&= -im\tau - z^2\bar{\tau}im + (1 + |\tau|^2)imz.
\end{aligned}
$$

Denoting $-im\tau := a$, $\quad (1 + |\tau|^2)im := b$ we obtain:

$$f(z) = a - \bar{a}z^2 + bz, \quad \text{Re}\,b = 0,$$

i.e., f is a complete vector field (see Proposition 3.5.1).

Therefore, we have to consider only the case when $\text{Re}\,p(z) > 0$.

Let us present $f \in \text{Hol}(\Delta)$ in the form:

$$f(z) = a - \bar{a}z^2 + z \cdot q(z), \tag{3.6.8}$$

where $a = f(0)$. Comparing (3.6.1) with (3.6.8) we have $f(0) = -\tau q(0)$ and

$$q(z) = (1 - \bar{\tau}z + |\tau|^2)q(z) - \frac{q(z) - q(0)}{z}\tau - z\overline{\tau q(0)}. \tag{3.6.9}$$

To proceed we need the following lemma which will be also useful in the sequel.

Lemma 3.6.1 (cf., [6]) *Let τ be in Δ and let p and q be those holomorphic functions in Δ which satisfy equation (3.6.9). Then $\text{Re}\,p(z) > 0$ if and only if $\text{Re}\,q(z) > 0$.*

Moreover, the values of p lie strictly inside $\Pi_+ = \{\text{Re}\,w > 0\}$ if and only if the values of q lie strictly inside Π_+.

Proof. First we note that assuming one of the functions p or q to be holomorphic in $\overline{\Delta}$ we have that the second one is holomorphic on $\overline{\Delta}$ too. Observe also that it is enough to prove our assertion under the above stronger assumption. Indeed, for the general case one can use an approximation argument: given p (or q, respectively) with $\operatorname{Re} p(z) > \varepsilon \geq 0$, set $p_n(z) = p(r_n z)$ for $r \in (0, 1)$, $r_n \to 1^-$.

So we assume that both these functions are holomorphic on $\overline{\Delta}$.

Then substituting $z = e^{i\theta}$, $\theta \in \mathbb{R}$, in (3.6.9) we calculate

$$
\begin{aligned}
\operatorname{Re} q(e^{i\theta}) &= \operatorname{Re}\{(1 + |\tau|^2 - \bar{\tau} e^{i\theta}) p(e^{i\theta}) - \tau e^{-i\theta} p(e^{i\theta})\} \\
&= (1 + |\tau|^2 - 2\operatorname{Re} \bar{\tau} e^{i\theta}) \operatorname{Re} p(e^{i\theta}) \\
&= |1 - \bar{\tau} e^{i\theta}|^2 \operatorname{Re} p(e^{i\theta}).
\end{aligned}
$$

Since by our assumptions, $\tau \in \Delta$, and the functions $\operatorname{Re} p$ and $\operatorname{Re} q$ are harmonic, we see that $\operatorname{Re} p(z) > \varepsilon \geq 0$ if and only if $\operatorname{Re} q(z) \geq \delta > 0$.

Moreover, if ε is positive, then δ can be chosen positive too, and conversely. □

Returning to the proof of the Proposition 3.6.1, we see that if $\tau \in \Delta$ then f defined by (3.6.1) admits representation (3.6.8) with $\operatorname{Re} q(z) > 0$, hence it is semi-complete (see Remark 3.5.2).

If τ in (3.6.1) belongs to $\partial \Delta$ we just apply again the following approximation argument. We choose any sequence $\tau_n \in \Delta$ such that $\tau_n \to \tau$ and set $f_n(z) = (z - \tau_n)(1 - z \bar{\tau}_n) p(z)$. It is obvious that $\{f_n\}_{n=1}^{\infty}$ converges to f uniformly on each compact subset of Δ. Since we already know that each f_n belongs to $\mathcal{G} \operatorname{Hol}(\Delta)$, we have that so does f, and we have completed the proof. □

Remark 3.6.1 Thus this proposition implies that *a semi-complete vector field* f *on* Δ *has at most one null point in* Δ. If such a point exists it must be τ in representation (3.6.1). For $\tau \in \Delta$ we will denote by $\mathcal{G} \operatorname{Hol}(\Delta, \tau)$ the class of functions $f \in \mathcal{G} \operatorname{Hol}(\Delta)$ with $f(\tau) = 0$. Thus,

$$
\mathcal{G} \operatorname{Hol}(\Delta, \tau) = \{f \in \operatorname{Hol}(\Delta, \mathbb{C}) : \ f(z) = (z - \tau)(1 - z\bar{\tau}) p(z), \ p \in \mathcal{P}\}.
$$

If τ *in this representation is a boundary point of* Δ, *then* f *is null point free. Moreover, it follows by the exponential formula that* τ *is, in fact, the sink point of the semigroup* $\{F_t\}_{t \geq 0}$ *generated by* f, *i.e., for each* $K > 0$ *the horocycle*

$$
D(\tau, K) = \left\{ z \in \Delta : \varphi_\tau(z) := \frac{|1 - z\bar{\tau}|^2}{1 - |z|^2} < K \right\}
$$

internally tangent to $\partial \Delta$ *at* τ, *is* F_t-*invariant, i.e.,*

$$
F_t(D(\tau, K)) \subseteq D(\tau, K).
$$

Also, it can be shown (see Section 4.6) that $\tau \in \partial \Delta$ is the *limit null point of* f in the following sense:

$$
\lim_{r \to 1^-} f(r\tau) = 0.
$$

However, it happens that $f \in \mathcal{G}\,\mathrm{Hol}(\Delta)$ may have more than one null point on the boundary of Δ (consider, for example, $f(z) = 1 - z^2$). So the question is which one of them is the sink point of the semigroup $\{F_t\}_{t \geq 0}$ generated by f. Another question relates to the asymptotic behavior of such a semigroup. In the next chapter we intend to answer these questions as much as to find the best rates of the exponential convergence.

At the end of the section we will add some preparatory material concerning this matter.

To clarify our further reasoning we first summarize briefly different characterizations of semi-complete vector fields.

Summary *Let $f \in \mathrm{Hol}(\Delta, \mathbb{C})$. The following are equivalent:*

(i) $f \in \mathcal{G}\,\mathrm{Hol}(\Delta)$, i.e., f is a semi-complete vector field;
(ii) for each $r > 0$ the mapping $J_r = (I + rf)^{-1}$ is a well defined holomorphic self-mapping of Δ;
(iii) f is ρ-monotone with respect to the Poincaré hyperbolic metric on Δ, i.e., for each pair $z, w \in \Delta$

$$\rho(z + rf(z), w + rf(w)) \geq \rho(z, w),$$

whenever $z + rf(z)$ and $w + rf(w)$ belong to Δ for some positive r;
(iv) f admits the following parametric representation

$$f(z) = a - \bar{a}z^2 + zq(z)$$

for some $a \in \mathbb{C}$ and $q \in \mathrm{Hol}(\Delta, \mathbb{C})$ with $\mathrm{Re}\, q(z) \geq 0$, $z \in \Delta$;
(v) f admits the Berkson–Porta parametric representation

$$f(z) = (z - \tau)(1 - z\bar{\tau})p(z)$$

for some $\tau \in \overline{\Delta}$ and $p \in \mathrm{Hol}(\Delta, \mathbb{C})$ with $\mathrm{Re}\, p(z) \geq 0$, $z \in \Delta$.

In addition, different flow invariance conditions given in terms inequalities are presented in sections 3.4 and 3.5. The simplest one can be formulated as follows
(vi) there exists a number $m \in \mathbb{R}$ (in fact, $m \leq 0$) such that

$$\mathrm{Re}\, f(z)\bar{z} \geq m(1 - |z|^2), \quad z \in \Delta.$$

In the study of asymptotic behavior of the semigroup generated by $f \in \mathcal{G}\,\mathrm{Hol}(\Delta)$ with an interior null point, a few stronger conditions than (i)–(vi) will be relevant.

Definition 3.6.1 (cf., [39]) *A function $f : \Delta \mapsto \mathbb{C}$ is said to be strongly ρ-monotone if for some $\varepsilon > 0$ and for each pair z, w in Δ there exists $\delta = \delta(z, w)$, such that*

$$\rho(z + rf(z), w + rf(w)) \geq (1 + r\varepsilon)\rho(z, w),$$

whenever $0 \leq r < \delta$.

Of course, a strongly ρ-monotone holomorphic function in Δ is semi-complete. Moreover, such a function must have a unique null point in Δ. Indeed, by definition we have that (at least for $r \geq 0$ small enough) the resolvent $J_r \in \mathrm{Hol}(\Delta)$ is a strict contraction with respect to the hyperbolic metric ρ in Δ:

$$\rho(J_r(z), J_r(w)) \leq \frac{1}{1 + r\varepsilon} \rho(z, w). \tag{3.6.10}$$

Thus J_r has a unique fixed point τ in Δ because of the Banach Fixed Point Principle. This point is a null point of f.

Further, putting $w = \tau$ in (3.6.10) and differentiating it with respect to r at the point $r = 0^+$ we obtain

$$(1 - |\tau|^2) \, \mathrm{Re} \, \frac{f(z)}{(z - \tau)(1 - z\bar{\tau})} \geq \varepsilon$$

or

$$\mathrm{Re} \, p(z) \geq \varepsilon (1 - |\tau|^2)^{-1} > 0,$$

where p is the factor in the Berkson–Porta representation (see (3.6.1)).

Now Lemma 3.6.2 enables us to conclude that $q(z)$ in representation (3.6.8) has a real part strictly separated from zero, i.e., $\mathrm{Re} \, q(z) > \varepsilon_1$ for some $\varepsilon_1 > 0$ and all $z \in \Delta$. In turn, the same formula (3.6.8) implies

$$\mathrm{Re} \, f(z)\bar{z} \geq -|a|(1 - |z|^2) + |z|^2 \varepsilon_1 > \varepsilon_2 > 0$$

for all z close enough to $\partial \Delta$, the boundary of Δ.

Thus we have proved the following assertion.

Proposition 3.6.2 *Let $f \in \mathrm{Hol}(\Delta, \mathbb{C})$ be strongly ρ-monotone in Δ. Then:*

(i) f admits the representation

$$f(z) = (z - \tau)(1 - z\bar{\tau})p(z)$$

with $\tau \in \Delta$ and $\mathrm{Re} \, p(z) > \varepsilon$ for some $\varepsilon > 0$;

(ii) there exist positive numbers δ and η such that

$$\mathrm{Re} \, f(z)\bar{z} \geq \eta > 0$$

for all z in the annulas $\{1 - \delta < |z| < 1\}$.

Again from Lemma 3.6.1 it can easily be seen that (ii) implies (i). Thus these conditions are equivalent. As a matter of fact, we will see below (Section 4.5) that (ii) implies the strong ρ-monotonicity of a function $f \in \mathrm{Hol}(\Delta, \mathbb{C})$.

Chapter 4

Asymptotic behavior of continuous flows

In this chapter we want to trace a connection of the iterating theory of functions in one complex variable and the asymptotic behavior of solutions of ordinary differential equations governed by evolution problems. Therefore our terminology is related to both these topics.

4.1 Stationary points of a flow on Δ

Quoting M. Abate [2], note that E. Vesentini seems to be the first person who suggested an analog of the Denjoy–Wolff Theorem for continuous time semigroups. In fact, in 1938 J. Wolff [158] himself initiated the consideration of dynamical systems determined by holomorphic functions. However, the first general continuous version of the Wolff–Denjoy Theory was given by E. Berkson and H. Porta [17] in their study of the eigenvalue problem for composition operators on Hardy spaces.

Definition 4.1.1 *A point $\zeta \in \overline{\Delta}$ is said to be a stationary point of a flow $S = \{F_t\}_{t>0} \subset \mathrm{Hol}(\Delta)$, if*

$$\lim_{r \to 1^-} F_t(r\zeta) = \zeta \tag{4.1.1}$$

for all $t > 0$.

In other words, $\zeta \in \Delta$ is a stationary point of S if it is a common fixed point of all $F_t \in S$.

Note that the family $S = \{F_t\}_{t \geq 0}$ is commuting, that is, $F_t \circ F_s = F_s \circ F_t = F_{t+s}$ for all $t, s \geq 0$. Hence, it follows by the Shield theorem [132] that if each F_t had been continuously extended to $\partial\Delta$ the boundary of Δ, then the stationary point set of S would not be empty.

As a matter of fact, it is enough to require the existence of an interior fixed point only for one $t > 0$ to ensure the existence of such a point for the whole semigroup. Indeed, if for at least one $t > 0$ the mapping $F_t \in S$ has an interior fixed point $\zeta \in \Delta$ then it is a unique fixed point for F_t, and for each $s \geq 0$ we have:

$$F_s(\zeta) = F_s(F_t(\zeta)) = F_t(F(\zeta)) = \zeta,$$

i.e., ζ is also a fixed point of $F_s \in S$, $s \geq 0$. Henceforth this fixed point is a unique stationary point of S.

Exercise 1. Show that if $\zeta \in \Delta$ is a stationary point for a semigroup $S = \{F_t\}_{t>0}$, $F_t \in \text{Hol}(\Delta)$, $t > 0$, then $(F_t)'(a) = e^{-at}$ is a contraction linear semigroup, i.e., $\text{Re}\, a \geq 0$.
Hint. Use the chain rule and the Schwarz–Pick lemma.

Naturally, the strategy now is to study the convergence of a semigroup to its stationary point. The foregoing result is the first step in the study of the asymptotic behavior of a flow in Δ.

Proposition 4.1.1 ([81]) *Let $S = \{F_t\}_{t>0} \subset \text{Hol}(\Delta)$ be a flow on Δ. Then this net converges uniformly on compact subsets of Δ to a holomorphic mapping $F \in \text{Hol}(\Delta, \mathbb{C})$ if and only if for at least one t_0 the sequence $\{F_{t_0n}\}_{n=0}^{\infty}$ converges uniformly on compact subsets of Δ. Moreover, if F_{t_0} is not the identity then F is a constant with the modulus less or equals to 1.*

Proof. The necessity is obvious. To prove the sufficiency we assume that F_t is not the identity for $t > 0$, otherwise the assertion is trivial. Then the limit:

$$\lim_{n\to\infty} F_{t_0n} = \lim_{n\to\infty} F_{t_0}^{(n)}$$

is a constant mapping, say $\zeta \ (= \zeta(t_0)) \in \overline{\Delta}$. If $\zeta \in \Delta$ then it follows by Corollary 1.3.2 that

$$\left| F_{t_0}'(\zeta) \right| < 1.$$

Consequently, the chain rule and the semigroup property imply that for all $t > 0$:

$$\left| F_t'(\zeta) \right| < 1 \tag{4.1.2}$$

(see also Exercise 1). Hence ζ is an attractive fixed point for each F_t, $t > 0$, i.e.,

$$\lim_{n\to\infty} F_t^{(n)}(z) = \zeta, \quad z \in \Delta. \tag{4.1.3}$$

If $\zeta \left(= \lim_{n\to\infty} F_{t_0}^{(n)} \right) \in \partial\Delta$, then F_{t_0} has no fixed point inside Δ. In this situation, as we mentioned above, each F_t, $t > 0$, must be fixed point free on Δ. Then the Denjoy–Wolff Theorem implies that for each integer $m > 0$ the sequence of iterates $\left\{ F_{t_0/m}^{(n)} \right\}_{n=0}^{\infty}$ converges uniformly on compact subset of Δ to a point $\zeta_m \in \Delta$.

But

$$\zeta_m = \lim_{n \to \infty} F_{t_0/m}^{(nm)}(z) = \lim_{n \to \infty} F_{t_0}^{(n)}(z) = \zeta,$$

and it follows that $\zeta_m = \zeta$ does not depend on m.

So, in both cases (either $\zeta \in \Delta$ or $\zeta \in \partial\Delta$) we have the following equality

$$\lim_{n \to \infty} F_{t_0/m}^{(n)}(z) = \zeta \qquad (4.1.4)$$

for all $z \in \Delta$, and each $m \in \mathbb{N}$.

Now we will show that (4.1.4) implies that for each $z \in \Delta$ the net F_t converges to ζ, as t tends to infinity. Indeed, for a given $\varepsilon > 0$ and $z \in \Delta$ one can choose $\delta > 0$ such that $|F_t(z) - F_t(w)| < \varepsilon/2$ for all $t > 0$, whenever $w \in \Delta$ and $|w - z| < \delta$. For such δ we take $m \in \mathbb{N}$ so large that $|F_s(z) - z| < \delta$ whenever $s \in [0, t_0/m]$. Finally, for such m and $t > 0$, large enough setting $s = [tm/t_0]$ we have by (4.1.4):

$$\left|F_{t_0 n/m}(z) - \zeta\right| = \left|F_{t_0/m}^{(n)}(z) - \zeta\right| < \frac{\varepsilon}{2}.$$

Noting that $s = t - t_0 n/m \in [0, t_0/m]$ and setting $w = F_s(z)$, we obtain for such $t > 0$:

$$\begin{aligned}
|F_t(z) - \zeta| &= \left|F_{t_0 n/m}(F_s(z)) - \zeta\right| \\
&\leq \left|F_{t_0 n/m}(F_h(z)) - F_{t_0 n/m}(z)\right| \\
&+ \left|F_{t_0 n/m}(z) - \zeta\right| \leq \frac{\varepsilon}{2} + \frac{\varepsilon}{2} = \varepsilon,
\end{aligned}$$

and we have completed the prof. $\square$

The result which we have established implies immediately a continuous analog of the Denjoy–Wolff Theorem.

Proposition 4.1.2 ([81]) *Let $S = \{F_t\}_{t \geq 0} \subset \mathrm{Hol}(\Delta)$ be a flow on Δ. If for at least one t_0 the mapping F_{t_0} is not the identity and is not an elliptic automorphism of Δ, then the net $\{F_t\}_{t \geq 0}$ converges to a constant $\zeta \in \overline{\Delta}$ as $t \to \infty$ uniformly on each compact subset of Δ.*

Remark 4.1.1 Since every continuous semigroup $S = \{F_t\}_{t \geq 0}$ of holomorphic self-mappings F_t of Δ is differentiable (by parameter $t \geq 0$), it is natural to describe its asymptotic behavior in terms of the generator $f = \lim_{t \to 0} \frac{1}{t}(I - F_t)$. This becomes more desirable when such a semigroup is not given explicitly, but is defined as the solution of the Cauchy problem:

$$\begin{cases}
\dfrac{\partial u(t, z)}{\partial t} + f\left(u(t, z)\right) = 0, \\[3mm]
u(0, z) = z \in \Delta.
\end{cases} \qquad (4.1.5)$$

We set here $F_t(z) = u(t, z)$.

Note also that if f is holomorphic in a neighborhood of the point $\zeta \in \overline{\Delta}$ then it follows by the uniqueness of the solution of the Cauchy problem that $f(\zeta) = 0$ if and only if ζ is a stationary point of $S = \{F_t\}_{t \geq 0}$. In particular, an interior null point of a semi-complete vector field is a stationary point of the generated semigroup. However, this fact is no longer true for a boundary null point. The following example shows that even a semi-complete vector field f has a continuous extension to $\overline{\Delta}$; it may have two null points in $\overline{\Delta}$ (one of them on $\partial \Delta$), while the semigroup generated by f has a unique stationary point in $\overline{\Delta}$ (which is the interior null point of f).

Example 1. Set $f(z) = z - 1 + \sqrt{1 - z}$. It is clear that $f(0) = f(1) = 0$. At the same time, solving the Cauchy problem (4.1.5) one can find the solution explicitly:

$$u(t, z) = 1 - \left[1 - e^{-t/2} + e^{-t/2}\sqrt{1 - z}\right]^2 .$$

Setting $F_t = u(t, \cdot)$, it is easy to verify that $F_t \in \text{Hol}(\Delta)$, hence f is semi-complete. But for all $t > 0$:

$$\lim_{r \to 1-} F_t(r) = 1 - \left[1 - e^{-t/2}\right]^2 < 1$$

and therefore $\zeta = 1$ is not a stationary point of $S = \{F_t\}_{t > 0}$.

Nevertheless, as we will see below (see Section 4.6), if f has no null point in Δ then it must have a boundary null point on $\partial \Delta$ which is an asymptotic limit of the semigroup generated by f.

4.2   Null points of complete vector fields

In this section we deal with a one-parameter group of automorphisms of Δ. Since the generator of a one-parameter group is a complete vector field, it is a polynomial at most of the second order, and hence holomorphic in $\mathbb{C}$. Now the assertion follows:

Proposition 4.2.1 *The stationary point set of a one-parameter group* $S = \{F_t\}_{t \in \mathbb{R}}$, $F_t \in \text{Aut}(\Delta)$, *has either one or two points in* $\overline{\Delta}$ *which are exactly the null points of the function* f:

$$f(z) = a + ibz - \bar{a}z^2, \qquad (4.2.1)$$

where

$$a = -\frac{\partial F_t(0)}{\partial t}\bigg|_{t=0} \quad and \quad b = \frac{1}{i}\frac{\partial^2 F_t(z)}{\partial t \partial z}\bigg|_{t=0, z=0}.$$

Exercise 1. Show directly that if $z_0 \neq 0$ is a solution of the equation

$$\bar{a}z^2 + ibz - a = 0, \quad b \in \mathbb{R}, \quad a \in \mathbb{C}, \tag{4.2.2}$$

then $z_1 = 1/\overline{z_0}$ is also a solution of this equation.

Thus equation (4.2.2) has at least one solution in $\overline{\Delta}$. If one of the solutions of (4.2.2) lies on $\partial\Delta$ the second one (if it exists) lies on $\partial\Delta$ also. Moreover, given a complete vector field f one can characterize the group of automorphisms generated by f.

Proposition 4.2.2 *Let* $f \in \text{aut}(\Delta)$ *be a complete vector field on* Δ *and let* $S = \{F_t\}$, $t \in \mathbb{R}$ *be the group of automorphisms generated by* f. *The following assertions hold:*

1) If $2|f(0)| < |f'(0)|$ *then* S *has a unique stationary point* z_0 *in* $\overline{\Delta}$ *which is actually in* Δ *and* S *consist of elliptic automorphisms of* Δ. *In this case* $F_t(z)$ *does not converge to* z_0 *for each* $z \in \Delta$, $z \neq z_0$.

2) If $2|f(0)| = |f'(0)|$ *then* S *has a unique stationary point* z_0 *in* $\overline{\Delta}$ *which actually lies on* $\partial\Delta$ *and* S *consists of parabolic automorphisms of* Δ. *In this case for all* $z \in \Delta$:

$$\lim_{t\to\infty} F_t(z) = z_0 \in \partial\Delta.$$

3) If $2|f(0)| < |f'(0)|$ *then* S *has exactly two different null points* z_1 *and* z_2 *in* $\overline{\Delta}$. *Both of them lie on* $\partial\Delta$, *and* S *consists of hyperbolic automorphisms of* Δ.

Consider in more details the group of hyperbolic automorphisms. In this case its generator f has the form:

$$f(z) = a + ibz - \bar{a}z^2, \tag{4.2.3}$$

with $a \in \Delta$, $b \in \mathbb{R}$, such that

$$2|a| > |b|. \tag{4.2.4}$$

Using (4.2.4), by direct calculations one obtains that the null points of f are

$$z_1 = \frac{2a}{\sqrt{4|a|^2 - b^2} + ib} \quad and \quad z_2 = \frac{-2a}{\sqrt{4|a|^2 - b^2} - ib}. \tag{4.2.5}$$

It is clear that $z_1 \neq z_2$ and $|z_1| = |z_2| = 1$.

Let now $S = \{F_t\}_{t\in\mathbb{R}}$ be the flow generated by f. Since for a fixed $t_0 \in \mathbb{R}^+$ the points z_1 and z_2 are fixed points of F_{t_0}, therefore one of them is a sink point for F_{t_0} and thus this point is the limit of the net $\{F_t\}$ when t goes to infinity.

It is easy to understand that the second null point is the sink point for the mapping $F_{t_0}^{-1} = F_{-t_0}$. Therefore it is the limit of the net $\{F_t\}$ when $t \to -\infty$.

Write F_t in the form:

$$F_t(z) = e^{i\varphi_t} \frac{z - a_t}{1 - \bar{a}_t z}, \tag{4.2.6}$$

where $a_t \in \Delta$, $\varphi_t \in \mathbb{R}$, $t \in \mathbb{R}$.

It is easily seen that for each $t \in \mathbb{R}$ and $z \in \Delta$:

$$\frac{F_t(z) - z_1}{F_t(z) - z_2} = \frac{1 - \bar{a}_t z_2}{1 - \bar{a}_t z_1} \cdot \frac{z - z_1}{z - z_2}.$$

This equation may be written in the form

$$L(F_t(z)) = \lambda_t L(z), \tag{4.2.7}$$

where L is the fractional linear transformation defined by

$$L(z) = \frac{z - z_1}{z - z_2} \tag{4.2.8}$$

and

$$\lambda_t = \frac{1 - \bar{a}_t z_2}{1 - \bar{a}_t z_1} = \frac{\bar{z}_2 - \bar{a}_t}{\bar{z}_1 - \bar{a}_t} \cdot \frac{z_2}{z_1}. \tag{4.2.9}$$

At the same time, since z_i, $i = 1, 2$ are fixed points of F_t we have by (4.2.6):

$$1 - \bar{a}_t z_i = \frac{e^{i\varphi_t}(z_i - a_t)}{z_i}.$$

It follows that

$$\lambda_t = \frac{1 - \bar{a}_t z_2}{1 - \bar{a}_t z_1} = \frac{z_2 - a_t}{z_1 - a_t} \cdot \frac{z_2}{z_1}. \tag{4.2.10}$$

In addition note, that (4.2.5) implies that

$$\frac{z_1}{z_2} = \frac{\bar{z}_2}{\bar{z}_1}.$$

(It is also clear because of the equality $z_i \bar{z}_i = 1$, $i = 1, 2$).

Comparing (4.2.9) and (4.2.10) we obtain that:

$$\lambda_t = \bar{\lambda}_t,$$

so the result that λ_t is real is established for all $t \in \mathbb{R}$.

Furthermore, it follows by the group property and (4.2.7) that

$$\lambda_{t+s} L(z) = L(F_{t+s}(z)) = L(F_t(F_s(z))) = \lambda_t L(F_s(z)) = \lambda_t \cdot \lambda_s L(z),$$

i.e.,

$$\lambda_{t+s} = \lambda_t \cdot \lambda_s, \quad \lambda_0 = 1.$$

So $\lambda_t = e^{kt}$ for some $k \in \mathbb{R}$.

If $k > 0$ we have again by (4.2.7):

$$F_t(z) = L^{-1}(e^{kt} L(z)),$$

or

$$F_t(z) = \frac{z_1 - e^{kt}\frac{z-z_1}{z-z_2}}{1 - e^{kt}\frac{z-z_1}{z-z_2}}. \tag{4.2.11}$$

Hence $\lim\limits_{t\to-\infty} F_t(z) = z_1$ while $\lim\limits_{t\to-\infty} F_t(z) = z_2$ for all $z \in \Delta$. In case of $k < 0$ we obtain similarly:

$$\lim_{t\to-\infty} F_t(z) = z_1$$

and

$$\lim_{t\to-\infty} F_t(z) = z_2.$$

Therefore we need to recognize k. Since $u(t,z) = F_t(z)$ satisfies the equation

$$\frac{\partial u(t,z)}{\partial t} + f(u(t,z)) = 0,$$

with the initial data:

$$u(0,z) = z,$$

we obtain by solving this equation:

$$\frac{F_t(z) - z_1}{F_t(z) - z_2} = e^{-\bar{a}(z_1 - z_2)\cdot t} \cdot \frac{z - z_1}{z - z_2}.$$

Comparing this formula with (4.2.7) we obtain that k must be $-\bar{a}(z_1 - z_2)$. By using (4.2.5) we calculate:

$$\begin{aligned} k &= -\bar{a}(z_1 - z_2) = -2|a|^2 \left(\frac{1}{\sqrt{4|a|^2 - b^2} + ib} + \frac{1}{\sqrt{4|a|^2 - b^2} - ib} \right) \\ &= -2|a|^2 \cdot \frac{\sqrt{4|a|^2 - b^2}}{4|a|^2} = -\sqrt{4|a|^2 - b^2} < 0. \end{aligned}$$

So k is negative in our case, and we have the second situation. Now we are ready to formulate our result.

Proposition 4.2.3 Let $f(z) = \bar{a}z^2 + ib - a$ such that $b \in \mathbb{R}$ and $2|a| > |b|$. Then f generates a flow of fractional linear transformations:

$$F_t(z) = e^{i\varphi_t} \frac{z - a_t}{1 - \bar{a}_t z},$$

which converges to

$$z_1 = -\frac{2a}{\sqrt{4|a|^2 - b^2} - ib}, \quad \text{as } t \to \infty,$$

and to

$$z_2 = \frac{2a}{\sqrt{4|a|^2 - b^2} + ib}, \quad \text{as } t \to -\infty.$$

In addition,

$$a_t = \frac{1 - e^{kt}}{\overline{z_2} - e^{kt}\overline{z_1}} \longrightarrow \frac{1}{\overline{z_2}} \in \partial\Delta,$$

where

$$k = -\sqrt{4|a|^2 - b^2}$$

and

$$\varphi_t = \arg \frac{z_1(1 - \overline{a_t}z_1)}{z_1 - a_t} = \arg \frac{z_2(1 - \overline{a_t}z_2)}{z_2 - a_t}.$$

Exercise 2. Let $f \in \mathrm{aut}(\Delta)$. Show that the fractional linear transformation G, defined by the formula:

$$G(z) = -\frac{z - c}{1 - z\overline{c}},$$

where $c = \dfrac{2f(0)}{f'(0)}$ has the same fixed points as the null points of f.

Exercise 3. Show that if $f \in \mathrm{aut}(\Delta)$ generates a one-parameter group of parabolic automorphisms of Δ then f has the form:

$$f(z) = \gamma(z - \tau)(\overline{\tau}z - 1),$$

where $\tau \in \partial\Delta$ and $\gamma \in \mathbb{C}$, $\mathrm{Re}\,\gamma = 0$. Find explicitly the group $S = \{F_t\}_{t\in\mathbb{R}}$ generated by f and show that

$$\lim_{t\to\infty} F_t(z) = \lim_{t\to-\infty} F_t(z) = \tau.$$

Exercise 4. Show that if $f \in \mathrm{aut}(\Delta)$ has a null point τ in Δ, i.e., f generates a one-parameter group $S = \{F_t\}_{t\in\mathbb{R}}$ of elliptic automorphisms, then $\mathrm{Re}\,f'(\tau) = 0$. Show that the group $S = \{F_t\}_{t\in\mathbb{R}}$ can be presented in the form:

$$F_t(z) = M_t\left(e^{at}M_\tau(z)\right),$$

where

$$M_\tau(z) = (z - \tau)(\overline{\tau}z - 1)^{-1}$$

and

$$\mathrm{Re}\,a = 0.$$

4.3 Embedding of discrete time group into a continuous flow

A classical problem of analysis is: given a self-mapping $F(z)$ ($F \in \text{Hol}(\Delta)$). Find a semigroup $\{F_t(z)\}_{t \geq 0} \subset \text{Hol}(\Delta)$ with $F_1(z) = F(z)$. In this case we say that F satisfies the embedding property on Δ. In general, however, this problem cannot be solved globally even if $F \in \text{Hol}(\Delta)$ is a fractional linear transformation (see Example 4.3.1).

Example 1 Consider the fractional linear transformation defined as follows:

$$F(z) = \frac{az}{1 - bz}$$

with $a = \frac{1}{3}\exp(i3.1)$ and $b = \frac{2}{3}$. Since $|a| + |b| = 1$ we have that F is a self-mapping of the unit disk Δ. Therefore its iterations $F^{(n)}$ map Δ into itself. Moreover, $F^{(n)}$ converge to zero uniformly on compact subsets of Δ as n goes to infinity. At the same time the semigroup

$$F_t := \frac{a^t z}{1 - \frac{b}{1-a} \cdot (1 - a^t)z}$$

is a continuous extension of the discrete time semigroup $\{F^{(n)}\}$ because of the equality $F = F_1$.

However, for some $t > 0$ (which is not integer) F_t does not map Δ into itself. If, for example, we choose $z = 0.99 + 0.1i \in \Delta$ then by calculation we obtain that $\left|F_{\frac{2}{3}}(z)\right| > 1$ (see Fig. 4.1).

Nevertheless, by using some geometric methods we will give in Section 5.9 a complete characterization of those fractional linear transformations of the unit disk into inself that satisfy the above embedding property. In this section we show how to solve this problem, when $F \in \text{Aut}(\Delta)$.

Suppose first that F is a hyperbolic automorphism of Δ, i.e., F has exactly two distinct fixed points z_1 and z_2 on the boundary $\partial\Delta$. As we saw above (see section 1.1.3), in this case if $F = r_\varphi \cdot m_\alpha$, then

$$F(z) = L^{-1}(\lambda F(z)), \tag{4.3.1}$$

where

$$\lambda = \frac{1 - \bar{\alpha}z_2}{1 - \bar{\alpha}z_1} \in \mathbb{R}^+ \setminus \{0, 1\}$$

and L is defined by

$$L(z) = \frac{z - z_1}{z - z_2}. \tag{4.3.2}$$

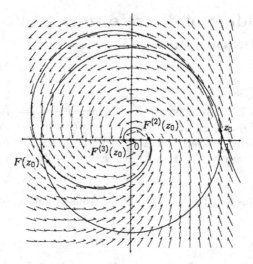

Figure 4.1: Fractional iterations of the self-mapping $F(z)$.

In addition, if z_1 is the sink point of F, then $0 < \lambda < 1$.
Denote

$$k = \log \lambda, \quad a = \frac{k}{z_2 - z_1}, \quad b = \frac{k}{i} \frac{z_1 + z_2}{z_z - z_2}, \quad \text{and} \quad c = \frac{z_1 \cdot z_2}{z_2 - z_1} k.$$

Since $|z_1| = |z_2| = 1$, and $z_1 \neq z_2$ it is easy to verify that $\bar{a} = -c$, and $b \in \mathbb{R}$.
Indeed,

$$\frac{\bar{a}}{c} = \frac{k}{z_1 - z_2} \cdot \frac{z_2 - z_1}{z_1 \cdot z_2 k} = \frac{z_2 - z_1}{(\overline{z_2} - \overline{z_1}) z_1 z_2} = \frac{z_2 - z_1}{z_1 |z_2|^2 - z_2 |z_1|^2} = -1.$$

In addition,

$$\begin{aligned} \operatorname{Re} ib &= k \operatorname{Re} \frac{z_1 + z_2}{z_1 - z_2} = \frac{k}{|z_1 - z_2|^2} \operatorname{Re}(z_1 + z_2)(\overline{z_1} - \overline{z_2}) = \\ &= \frac{k}{|z_1 - z_2|^2} (\operatorname{Re} |\overline{z_1}|^2 + z_2 \overline{z_1} - \overline{z_2} z_1 - |z_2|^2) = 0, \end{aligned}$$

i.e., $b \in \mathbb{R}$. Therefore the mapping f:

$$f(z) = az^2 + ibz + c$$

is a complete vector field. It follows by the Vietta formulas that z_1 and z_2 are
the roots of the equation $f(z) = 0$. Hence as above (see section 4.2) the group of
automorphisms generated by f has a form:

$$F_t(z) = L^{-1}(e^{kt} L(z)),$$

where L is defined by (4.4.2).

For $t = 1$ we have:

$$F_1(z) = L^{-1}(e^{\log \lambda} L(z)) = F(z)$$

because of (4.4.1). This gives us the desirable embedding.

As a matter of fact, we have proved somewhat more:

Proposition 4.3.1 *Let $z_1 \neq z_2$ are two arbitrary points on the boundary $\partial\Delta$ of the unit disk Δ. Then for each $k \in \mathbb{R}$, $k \neq 0$, there is a one-parameter group $S = \{F_t\}$, $t \in \mathbb{R}$, such that the following assertions hold:*

(a) $F_t(z_i) = z_i$, $i = 1, 2$.

(b) $\dfrac{dF_t(z)}{dt}\big|_{t=0} = -(az^2 + ibz - \bar{a})$ where $a = \dfrac{k}{z_2 - z_1}$, $b = \dfrac{k}{i}\dfrac{z_1 + z_2}{z_2 - z_1}$.

(c) If $k < 0$ then z_1 is a sink point of $\{F_t\}_{t>0}$; if $k > 0$ then z_2 is a sink point of $\{F_t\}_{t>0}$.

(d) In particular, if z_1 and z_2 are fixed points of a hyperbolic automorphism

$$F(z) = e^{i\varphi} \frac{z + \alpha}{1 + \bar{\alpha}z}, \quad \alpha \in \Delta, \quad \varphi \in \mathbb{R},$$

then for

$$k = \ln \frac{1 - \bar{\alpha}z_2}{1 - \bar{\alpha}z_1}$$

we have the equality:

$$F_1(z) = F(z), \quad z \in \Delta.$$

Now we turn to a parabolic automorphism $F \in \text{Aut}(\Delta)$. In this case F has a unique fixed point $z_0 \in \partial\Delta$ which is a sink point of F. Moreover, as we already know F satisfies the equation:

$$\frac{z_0}{F(z) - z_0} = \frac{e^{i\varphi} - 1}{e^{i\varphi} + 1} + \frac{z_0}{z - z_0} \tag{4.3.3}$$

for some $\varphi \in \mathbb{R}$, $\varphi \neq \pi + 2n\pi$ (see section 1.3.1).

It is clear that in such a situation we must solve the following Cauchy problem:

$$\begin{cases} \dfrac{\partial u(t, z)}{\partial t} + a(z - z_0)^2 = 0, \\[2mm] u(0, z) = z \in \Delta, \end{cases} \tag{4.3.4}$$

with suitable $a \in \mathbb{C}$ in order to find a one-parameter group $\{F_t(\cdot)\}$, $F_t = u(t, \cdot)$, such that:

$$F_1(z) = F(z). \tag{4.3.5}$$

Solving (4.3.4) we have that F_t satisfies the equation:

$$\frac{1}{F_t(z) - z_0} = at + \frac{1}{z - z_0}.$$

Comparing the latter equality with (4.3.3) and (4.3.5) we obtain:

$$a = \frac{e^{i\varphi} - 1}{e^{i\varphi} + 1} \cdot \frac{1}{z_0}. \tag{4.3.6}$$

Now we only need to check that the mapping f defined by:

$$f(z) = a(z - z_0)^2 \tag{4.3.7}$$

is a complete vector field. In fact, substitute (4.3.6) in (4.3.7) we obtain:

$$f(z) = a\, z^2 + ibz + c,$$

where

$$c = az_0^2 = \frac{e^{i\varphi} - 1}{e^{i\varphi} + 1} \cdot z_0 = -\bar{a}$$

and

$$b = \frac{-2}{i} \cdot \frac{e^{i\varphi} - 1}{e^{i\varphi} + 1} = -\frac{4\,\mathrm{Im}\,e^{i\varphi}}{|e^{i\varphi} + 1|^2} \in \mathbb{R}.$$

Thus we have proved the following assertion:

Proposition 4.3.2 *Let z_0 and $\tau \neq -1$ be two unimodular points. Then the mapping f defined by (4.3.7) with*

$$a = \frac{\tau - 1}{\tau + 1}\overline{z_0}$$

is a generator of the one-parameter group $S = \{F_t\}_{t \in \mathbb{R}} \in \mathrm{Aut}(\Delta)$, such that each F_t is a parabolic automorphism of Δ, and the point $z_0 \in \partial\Delta$ is a sink point of S.

If, in particular, $F \in \mathrm{Aut}(\Delta)$, is a parabolic automorphism of Δ:

$$F(z) = e^{i\varphi} \frac{z + \alpha}{1 + z\bar{\alpha}}$$

such that $F(z_0) = z_0$, then for $\tau = e^{i\varphi}$ we have: $F_1(z) = F(z)$.

Finally, we consider the simplest situation where $F \in \mathrm{Aut}(\Delta)$ is an elliptic automorphism of Δ, i.e., F has a unique fixed point $z_0 \in \Delta$.

In this case F can be presented as:

$$F = m_{z_0} \circ r_\varphi \circ m_{-z_0}, \tag{4.3.8}$$

where

$$m_{z_0}(z) = \frac{z + z_0}{1 + \overline{z_0}z}$$

is a Möbius transformation of Δ, $m_{z_0}(0) = z_0$, and r_φ is a rotation around zero, i.e.,

$$r_\varphi(z) = e^{i\varphi}z, \quad \varphi \in \mathbb{R}, \quad \varphi \neq 2\pi k, \quad k = 0, 1, 2, \ldots .$$

Setting in this case

$$r_{\varphi_t}(z) = e^{i\varphi t}z$$

and

$$F_t(z) = m_{z_0} \circ r_{\varphi_t} \circ m_{-z_0}, \quad t \in \mathbb{R}, \tag{4.3.9}$$

we have that $S = \{F_t\}$, $t \in \mathbb{R}$ is a one-parameter group of elliptic automorphisms of Δ such that $F_t(z_0) = z_0$ for all $t \in \mathbb{R}$ and $F_t(z) = F(z)$, $z \in \Delta$.

Exercise 1. Prove that group (4.3.9) is generated by the mapping f:

$$f(z) = \frac{-i\varphi}{1 - |z_0|^2}(z - z_0)(1 - \overline{z_0}z). \tag{4.3.10}$$

Exercise 2. Prove directly that the mapping f defined by (4.3.10) is a complete vector field.

Finally, we formulate the result:

Proposition 4.3.3 *Let $z_0 \in \Delta$ and $\varphi \in \mathbb{R}$, $\varphi = 2\pi k$, $k = 0, 1, 2 \ldots$. Then the mapping f defined by (4.3.10) generates the one-parameter group $S = \{F_t\}$, $t \in \mathbb{R}$ of elliptic automorphisms defined by formula (4.3.9). In particular, if $F \in \mathrm{Aut}(\Delta)$ is an elliptic automorphism, defined by (4.3.8), then $F_1(z) = F(z)$, $z \in \Delta$.*

4.4 Rates of convergence of a flow with an interior stationary point

Let $f \in \mathrm{Hol}(\Delta, \mathbb{C})$ be a semi-complete vector field on Δ, and let us assume that $\mathrm{Null} f \cap \Delta \neq \emptyset$. Actually, it means that if f is not zero identically, then f has exactly one null point in Δ (otherwise the semigroup $S = \{F_t\}_{t \geq 0}$ generated by f should have more than one stationary point in Δ, and a contradiction results). A standard question of dynamical systems analysis is: given a vector field f, describe the asymptotic behavior of the flow defined by the evolution equation $du/dt + f(u) = 0$, $u(0) = z$, in a neighborhood of its singular point τ ($f(\tau) = 0$).

In this section we intend to answer this question for a semi-complete vector field, as much as to find global rates of convergence of the generated flow to its interior stationary point.

For our purpose we need the following definition.

Definition 4.4.1 ([40]) *A function $f \in \mathcal{G} \mathrm{Hol}(\Delta)$ is said to be a strongly semi-complete vector field if it has a unique null point τ in Δ which is a uniformly attractive stationary point for the flow $S = \{F_t\}_{t \geq 0}$ generated by f, that is, the net $\{F_t\}_{t \geq 0}$ converges to τ uniformly on each compact subset of Δ as $t \to \infty$.*

We begin with establishing a simple assertion which is a (nonlinear) analog of the Lyapunov stability theorem.

Proposition 4.4.1 *Let $f \in \text{Hol}(\Delta, \mathbb{C})$ be a semi-complete vector field with $f(\tau) = 0$ for some $\tau \in \Delta$. Then:*

(i) $\text{Re } f'(\tau) \geq 0$;

(ii) $\text{Re } f'(\tau) > 0$ if and only if f is strongly semi-complete.

Proof. Let $S = \{F_t\}_{t \geq 0}$ be the flow generated by f. Since $F_t(\tau) = \tau$ for all $t \geq 0$, it follows by the chain rule, that $(F_t)'(\tau) = A_t$ satisfies the semigroup property $A_{t+s} = A_t \cdot A_s$, $A_0 = 1$, hence $A_t = e^{kt}$, with $k = f'(\tau) \in \mathbb{C}$.

In addition, it follows by Corollary 1.1.1 that

$$\left| e^{-kt} \right| \leq 1.$$

But this inequality holds if and only if $\text{Re } k = \text{Re } f'(\tau) \geq 0$ and assertion (i) follows. Furthermore, $\text{Re } k = \text{Re } f'(\tau) > 0$ if and only if for all $t \geq 0$

$$\left| e^{-kt} \right| = |(F_t)'(\tau)| < 1,$$

and then assertion (ii) is a consequence of Proposition 4.1.1. and Corollary 3.1.1.
□

Exercise 1. Show that if $\text{Re } f'(\tau) = 0$ then the semigroup $S = \{F_t\}_{t \geq 0}$ is actually a group of elliptic automorphisms.

Thus, if $f \in \mathcal{G} \text{Hol}(\Delta)$ is not a complete vector field of the type

$$f(z) = (z - \tau)(1 - \bar{\tau}z)a$$

with $\tau \in \Delta$ and $\text{Re } a = 0$, then the semigroup $S = \{F_t\}_{t \geq 0}$ generated by f does not consist of elliptic automorphisms and converges uniformly on each compact subset of Δ to a point in $\overline{\Delta}$. If this point belongs to Δ, then it is a unique uniformly attractive stationary point of the semigroup, and the question of finding a rate of convergence arises.

Remark 4.4.1 We already known that for a holomorphic function f on Δ the property of being semi-complete is equivalent to the property of being monotone with respect to the Poincaré metric ρ on the unit disk. Also, it can easily be seen that if f is strongly ρ-monotone then it is strongly semi-complete. For this case one can obtain a rate of convergence in terms of the hyperbolic metric on Δ (see Section 4.5). The converse, however, does not hold in general: a strongly semi-complete vector field is not necessarily strongly ρ-monotone.

Therefore in this section we will establish some rates of convergence in terms of the Euclidean and pseudo-hyperbolic distance on Δ.

First we consider the case when $f(0) = 0$, i.e., zero is a stationary point of the semigroup $S = \{F_t\}_{t \geq 0}$.

Proposition 4.4.2 (see [58] and [109]) *Let $f \in \mathcal{G}\,\mathrm{Hol}(\Delta)$ be a strongly semi-complete vector field with $f(0) = 0$ and $\lambda = \mathrm{Re}\, f'(0) > 0$, and let $S = \{F_t\}_{t \geq 0}$ be the semigroup generated by f. Then there exists $c \in [0,1]$ such that for all $z \in \Delta$ and $t \geq 0$ the following estimates hold:*

(a) $|F_t(z)| \leq |z| \cdot \exp\left(-\lambda \dfrac{1 - c|z|}{1 + c|z|} t\right);$

(b) $\dfrac{|F_t(z)|}{(1 - c|F_t(z)|)^2} \leq \exp(-\lambda t) \dfrac{|z|}{(1 - c|z|)^2}.$

Proof. The procedure that is to follow is based on general flow invariance conditions established in sections 3.4 and 3.5. We start with the consideration of an auxiliary Cauchy problem:

$$
\begin{cases}
\dfrac{\partial u(t, z)}{\partial t} + p \cdot u(t, z) \cdot \dfrac{1 - cu(t, z)}{1 + u(t, z)} = 0, \\[2mm]
u(0, z) = z, \quad z \in \Delta.
\end{cases}
\tag{4.4.1}
$$

Noting that the function $\alpha(z) = p \cdot z \dfrac{1 - z}{1 + z}$ satisfies the condition

$$
\mathrm{Re}\, \alpha(z)\bar{z} = |z|^2 p\, \mathrm{Re}\, \frac{1 - z}{1 + z} \geq 0
$$

for all $z \in \Delta$, we conclude that $\alpha \in \mathcal{G}\,\mathrm{Hol}(\Delta)$ and the Cauchy problem above has a unique solution $u(t, z) \in \Delta$ for all $z \in \Delta$ and $t \geq 0$. In addition, $u(t, 0) = 0$ for all $t \geq 0$, because of $\alpha(0) = 0$. Then by the Schwarz Lemma we obtain that

$$
|u(t, z)| \leq |z|, \quad z \in \Delta.
\tag{4.4.2}
$$

Furthermore, since the function $\alpha(s) = ps\dfrac{1 - s}{1 + s}$, defined on the interval $[0, 1]$ is real we have by the uniqueness property that the differential equation:

$$
\frac{dv}{dt} + \alpha(v) = 0
\tag{4.4.3}
$$

with the initial data:

$$
v(0) = s
\tag{4.4.4}
$$

has a unique solution $v(t, s)$ for all $t \geq 0$ and $s \in [0, 1)$ which coincides with $u(t, s)$ restricted on the real interval $[0, 1)$. Consequently, by (4.4.2) we obtain:

$$
v(t, s) < s
\tag{4.4.5}
$$

for all $t \geq 0$ and $s \in [0, 1)$. Also, it follows from (4.4.3) that:

$$
\frac{dv}{v} = -p\frac{1 - v}{1 + v}dt.
$$

Hence, by (4.4.4):

$$v = s \exp\left(\int -p\frac{1-v}{1+v} dt \right).$$

Finally, inequality (4.4.5) implies the estimate:

$$v(s,t) \le s \exp\left(-p\frac{1-s}{1+s}t \right). \qquad (4.4.6)$$

Recall that by Proposition 3.4.5 $f \in \mathcal{G}\,\mathrm{Hol}(\Delta)$ with $f(0) = 0$ satisfies the inequality:

$$\mathrm{Re}\, f(z)\bar{z} \ge \mathrm{Re}\, f'(0)|z|^2\frac{1-|z|}{1+|z|} = \alpha(|z|) \cdot |z|, \quad z \in \Delta.$$

Then by Lemma 3.4.1 and (4.4.6) we have:

$$|F_t(z)| \le u(t,|z|) \le |z| \cdot \exp\left(-\mathrm{Re}\, f'(0)\frac{1-|z|}{1+|z|}t \right).$$

This proves (a).

To prove (b) we return to (4.4.3) and rewrite it in the form:

$$\frac{1+v}{1-v}\cdot\frac{1}{v}dv = -pdt$$

(we substitute $\alpha(v) = vp\dfrac{1+v}{1-v}$ into this equation). Integrating the latter equality in $[0,t]$ we obtain:

$$\frac{v(s,t)}{(1-v(s,t))^2} = e^{-pt}\frac{s}{(1-s)^2}, \quad s \in [0,1).$$

Again, setting $s = |z|$ and using the relation

$$|F(t,|z|)| \le v(s,t),$$

we obtain (b). $\square$

Note that estimate (a) with $c = 1$ is due to K. Gurganus, whilst estimate (b) was established by T. Poreda.

Now we are prepared to consider a more general case, when $f \in \mathcal{G}\,\mathrm{Hol}(\Delta)$, with $f(\tau) = 0$ for some $\tau \in \Delta$, which is not necessarily zero.

Proposition 4.4.3 *Let Δ equipped with the pseudo-hyperbolic distance*

$$d(z,w) = \left| \frac{w-z}{1-\bar{w}z} \right|,$$

and let $S_f = \{F_t\}_{t\ge0}$ be the semigroup generated by $f \in \mathcal{G}\,\mathrm{Hol}(\Delta)$. Then for some $\tau \in \Delta$ the following conditions are equivalent:

(i) $f(\tau) = 0$ with $\operatorname{Re} f'(\tau) \geq p \geq 0$;
(ii) for some $c \in [0,1]$:

$$d\left(F_t(z), \tau\right) \leq d(z, \tau) \exp\left(-p \frac{1 - cd(z,\tau)}{1 + cd(z,\tau)} t\right)$$

for all $t \geq 0$ and $z \in \Delta$.

In addition, if f is bounded and strongly p-monotone then the number c in (ii) can be chosen as strictly less than 1.

Proof. Define the Möbius transformation $M_\tau : \Delta \mapsto \Delta$ by:

$$M_\tau(z) = \frac{\tau - z}{1 - \bar{\tau}z}, \quad z \in \Delta. \tag{4.4.7}$$

It is clear that

$$M_\tau(0) = \tau, \quad M_\tau(\tau) = 0, \tag{4.4.8}$$

and it follows that M_τ is an involution, i.e.,

$$M_\tau = M_\tau^{-1}. \tag{4.4.9}$$

Consider the family $\{G_t\}_{t \geq 0}$ defined as follows:

$$G_t = M_\tau \circ F_t \circ M_\tau. \tag{4.4.10}$$

It is obvious that $\{G_t\}$ is also a semigroup of holomorphic self-mappings of Δ, and for all $t \geq 0$ we obtain

$$G_t(0) = 0, \quad t \geq 0, \tag{4.4.11}$$

since $F_t(\tau) = \tau$, for all $t \geq 0$.

Let g be the generator of the semigroup $\{G_t\}_{t \geq 0}$, i.e.,

$$g(z) = \frac{-dG_t(z)}{dt}\bigg|_{t=0+}. \tag{4.4.12}$$

By direct calculations we obtain:

$$\begin{aligned}
\frac{d}{dt}(G_t(z)) &= \frac{d}{dt}\left(M_\tau(F_t(M_\tau(z)))\right) = \frac{dM_\tau}{dz}\left(F_t(M_\tau(z))\right) \cdot \frac{dF_t}{dt}(M_\tau(z)) \\
&= \frac{dM_\tau}{dz}(F_t(M_\tau(z))) \cdot (-f(F_t(M_\tau(z)))), \quad z \in \Delta.
\end{aligned}$$

Hence by (4.4.12)

$$g(z) = \frac{dM_\tau}{dz}(M_\tau(z)) \cdot f(M_\tau(z)), \quad z \in \Delta. \tag{4.4.13}$$

Note that the latter formula expresses the direct connection between generators f and g. Also, (4.4.8) and (4.4.13) imply:

$$g(0) = 0, \tag{4.4.14}$$

which is equivalent to (4.4.11).

In turn, differentiating (4.4.13) we calculate

$$g'(z) = \frac{d^2 M_\tau}{dz^2}(M_\tau(z)) \cdot \frac{dM_\tau}{dz}(z) \cdot f(M_\tau(z))$$
$$+ \frac{dM_\tau}{dz}(M_\tau(z)) \cdot \frac{df}{dz}(M_\tau(z)) \cdot \frac{dM_\tau}{dz}(z). \tag{4.4.15}$$

Substituting here $z = 0$ and using that $f(\tau) = 0$ we obtain:

$$g'(0) = \frac{dM_\tau}{dz}(\tau) \cdot \frac{df}{dz}(\tau) \cdot \frac{dM_\tau}{dz}(0). \tag{4.4.16}$$

Since

$$\frac{dM_\tau}{dz} = \frac{d}{dz}\left(\frac{\tau - z}{1 - \bar{\tau}z}\right) = \frac{|\tau|^2 - 1}{(1 - \bar{\tau}z)^2}$$

relation (4.4.16) becomes:

$$g'(0) = \frac{1}{|t|^2 - 1} f'(\tau) \cdot (|\tau|^2 - 1) = f'(\tau). \tag{4.4.17}$$

So by Proposition 4.4.2 (a) we obtain:

$$|G_t(z)| \le |z| \exp\left(-\operatorname{Re} g'(0) \cdot \frac{1 - |z|}{1 + |z|} \cdot t\right),$$

and using (4.4.10) and (4.4.17) we obtain from the latter inequality:

$$|(M_\tau \circ F_t \circ M_\tau)(z)| \le |z| \exp\left(-\operatorname{Re} f'(\tau) \cdot \frac{1 - |z|}{1 + |z|} \cdot t\right). \tag{4.4.18}$$

For arbitrary w, setting $z = M_\tau(w)$ in (4.4.18) we obtain:

$$|(M_\tau \circ F_t)(w)| \le |M_\tau(w)| \exp\left(-\operatorname{Re} f'(\tau) \cdot \frac{1 - |M_\tau(w)|}{1 + |M_\tau(w)|} \cdot t\right). \tag{4.4.19}$$

Rewriting (4.4.19) in the form:

$$\left|\frac{\tau - F_t(w)}{1 - F_t(w)\bar{\tau}}\right| \le \left|\frac{\tau - w}{1 - \bar{\tau}w}\right| \exp\left(-\operatorname{Re} f'(\tau) \cdot \frac{1 - \left|\frac{\tau - w}{1 - \bar{\tau}w}\right|}{1 + \left|\frac{\tau - w}{1 - \bar{\tau}w}\right|} \cdot t\right), \tag{4.4.20}$$

we obtain (ii).

Conversely, assuming that (ii) holds we first observe that $f(\tau)$ should be zero. Of course, this is an immediate consequence of (ii) if $p > 0$, hence $\operatorname{Re} f'(\tau) > 0$. Indeed, in this case

$$\lim_{t \to \infty} F_t(z) = \tau,$$

so, τ is the stationary point of $S_f = \{F_t\}_{t \ge 0}$, whence, a null point of f. But even if $\operatorname{Re} f'(0) = 0$, condition (ii) means that each pseudo-hyperbolic ball $\Omega_r(\tau)$ (see Section 1.3) is F_t-invariant for all $t \ge 0$, and the same conclusion follows.

To arrive at the second condition of (i) we just need to differentiate (ii) at $t = 0^+$ to obtain:

$$(1 - |\tau|^2) \operatorname{Re} \frac{f(z)}{(z - \tau)(1 - z\bar{\tau})} \geq p \frac{1 - \left| \frac{\tau - z}{1 - \bar{\tau}z} \right|}{1 + \left| \frac{\tau - z}{1 - \bar{\tau}z} \right|}.$$

Letting $z \to \tau$ we obtain immediately that $\operatorname{Re} f'(\tau) \geq p$ as desired. Finely, note that our last assertion follows from the strong Harnack inequality (see Section 3.5 and Proposition 3.6.2). $\square$

Remark 4.4.2 Since $|F_t(w)| < 1$ one may obtain from (4.4.20) (setting $c = 1$) the following estimate

$$|F_t(w) - \tau|$$
$$\leq |w - \tau| \frac{1 + |\tau|}{1 - |\tau| |w|} \exp\left(-\operatorname{Re} f'(\tau) \cdot \frac{|1 - \bar{\tau}w| - |\tau - w|}{|1 - \bar{\tau}w| + |\tau - w|} t \right). \qquad (4.4.21)$$

This formula gives an estimated rate of convergence of the semigroup $\{F_t\}_{t>0}$ to its stationary point τ in the usual Euclidean metric in $\mathbb{C}$. As a matter of fact, this estimate can be slightly improved by using the following considerations.

Return to formula (4.4.19). Denoting $F_t(w) = J$ and $M_\tau(J) = u$ we have $M_\tau(u) = J$ and $M_\tau(0) = \tau$.

Now, it follows by Corollary 1.1.3 of the Schwartz Lemma (see section 1.1.1), that:

$$|M_\tau(u) - M_\tau(0)| \leq |u| \frac{1 - |M_\tau(0)|^2}{1 - |M_\tau(0)||u|} = |u| \frac{1 - |\tau|^2}{1 - |\tau| |u|}$$

or

$$|F_t(w) - \tau|$$
$$\leq \frac{|\tau - w|}{|1 - \bar{\tau}w|} \cdot \frac{(1 - |\tau|^2) \cdot \exp\left(-\operatorname{Re} f'(\tau) \cdot \frac{|1 - \bar{\tau}w| - |\tau - w|}{|1 - \bar{\tau}w| + |\tau - w|} \cdot t \right)}{1 - |\tau| \cdot \exp\left(-\operatorname{Re} f'(\tau) \cdot \frac{|1 - \bar{\tau}w| - |\tau - w|}{|1 - \bar{\tau}w| + |\tau - w|} \cdot t \right)}. \qquad (4.4.22)$$

It is clear that (4.4.22) implies (4.4.21).

Remark 4.4.3 Note that the above proposition may be considered as an analogy of the Schwarz–Pick inequality. Moreover, for the continuous case this inequality has a more precise form because of the exponential factor

$$\exp\left(-p \frac{1 - d(z, \tau)}{1 + d(z, \tau)} t \right).$$

However, the property of this factor depending also on $z \in \Delta$ is sometimes inconvenient for applications. In some situations it would be preferable to find a metric on Δ in which such a factor becomes uniform. In addition, as in the discrete time case, condition (ii) in our proposition cannot be useful in the study of the boundary asymptotic behavior of flows with no stationary point inside Δ. Therefore we will treat the last question separately in Section 4.6 by using a continuous extension of the Julia–Wolff–Caratheodory Theorem. However, it turns out that one can define a non-Euclidean distance on Δ (which, in fact, is not a metric on Δ, but induces the original topology of Δ) such that the rates of convergence for interior and boundary points have some unified form. This makes it possible to study the dynamic behavior of evolution equations when their stationary points approach the boundary. Moreover, we will show in Section 4.6 that in terms of such a distance, the question about a uniform rate of convergence of the exponential type may be solvable too.

Finally, we note that a deficiency of the rates of convergence given by formulas (4.4.20) and (4.4.21) is the presence there the value of the derivative f' at the null point τ of a generator f. In fact, if f has a complicated form even the existence of its null point is unclear *a priori*. Therefore it would be nice to establish a rate of convergence by using another condition which will guarantee also the existence of an interior null point of a semi-complete vector field. The next section is devoted to this matter.

Remark 4.4.4 From the geometrical function theory point of view there is a need to point out a lower bound estimate of the flow behavior. Exactly as in Proposition 4.4.2, by using the left side inequality of (3.5.18) and Lemma 3.5.1 one again obtains: *if $f(0) = 0$ and $\operatorname{Re} f'(0) = \lambda \geq 0$, then for some $c \in [0, 1]$*

$$(a') \quad |F_t(z)| \geq |z| \exp(-\lambda \frac{1 + c|z|}{1 - c|z|} t);$$

$$(b') \quad \frac{|F_t(z)|}{(1 - c|F_t(z)|)^2} \geq \exp(-\lambda t) \frac{|z|}{(1 + c|z|)^2}.$$

Moreover, if f is bounded strongly ρ-monotone then c can be chosen strictly less then 1.

4.5   A rate of convergence in terms of the Poincaré metric

In this section we will give several sufficient conditions for $f \in \operatorname{Hol}(\Delta, \mathbb{C})$ to be strongly ρ-monotone, hence strongly semi-complete on the open unit disk Δ and obtain rates of convergence for the semigroups generated by such functions in terms of the Poincaré hyperbolic metric ρ on Δ.

Proposition 4.5.1 ([40]) *Let $f \in \mathrm{Hol}(\Delta, \mathbb{C})$ satisfy the condition:*

$$\mathrm{Re}\, f(z)\bar{z} \geq \alpha(|z|)|z|, \qquad z \in \Delta \qquad (4.5.1)$$

for some real continuous function $\alpha(l)$ on the interval $[0, 1]$, such that:

$$\alpha(1) = \alpha > 0. \qquad (4.5.2)$$

Then f is strongly ρ-monotone, hence a strongly semi-complete vector field on Δ. Thus f has a unique null point $\tau \in \Delta$ which is uniformly attractive for the semigroup $S = \{F_t\}$, $t \geq 0$, generated by f. Moreover, for each pair $z, w \in \Delta$

$$\rho(F_t(z), F_t(w)) \leq \exp\left\{-\frac{\alpha t}{2}\right\}\rho(z, w). \qquad (4.5.3)$$

In particular,

$$\rho(F_t(z), \tau) \leq \exp\left\{-\frac{\alpha t}{2}\right\}\rho(z, w). \qquad (4.5.4)$$

Proof. Consider the mapping $g_s : \Delta \mapsto \mathbb{C}$ defined as follows:

$$g_s(z) = z + sf(z) - w, \qquad \text{where} \quad w \in \Delta, \ s \geq 0.$$

Let Δ_r be the disk centered at zero with radius $r : 0 \leq r < 1$. For all $z \in \partial\Delta_r = \{z \in \mathbb{C} : |z| = r\}$ we have by (4.5.1):

$$
\begin{aligned}
\mathrm{Re}\, g_s(z)\bar{z} &= |z|^2 + s\,\mathrm{Re}\, f(z)\bar{z} - \mathrm{Re}\, w\bar{z} \geq r^2 + s\alpha(r)r - r|w| \\
&= r\,(r + s\alpha(r) - |w|)\,. \qquad (4.5.5)
\end{aligned}
$$

Since $\alpha(1) > 0$ it follows that for $s > 0$ small enough the equation:

$$\varphi_s(r) = r + s\alpha(r) = 1 \qquad (4.5.6)$$

has a solution $r_s \in (0, 1)$.

In fact, $\varphi_s(0) = s\alpha(0) \leq 1$ for $0 < s < 1/\alpha(0)$ and $\varphi_s(1) = 1 + s\alpha(1) \geq 1$. It follows by (4.5.5) that for such fixed n and $z \in \partial\Delta_{r_n}$,

$$\mathrm{Re}\, g_s(z) \cdot \bar{z} \geq 0.$$

The latter inequality implies that the equation:

$$g_s(z) = z + sf(z) - w = 0$$

has a unique solution $z = J_s(w) := (I + sf)^{-1}(w) \in \Delta_{r_s}$ for each $w \in \Delta$.

In other words, the resolvent mapping J_s maps Δ into Δ_{r_s}.

This means that J_s has an interior fixed point $\tau \in \Delta$ which is also a unique null point of f in Δ.

Furthermore, we consider the function $h(z) = h_s(z)$ defined as follows:

$$h_s(z) = J_s(z) + s\frac{\alpha(r_s)}{2}\left(J_s(z) - J_s(w)\right),$$

where w is an element of Δ, and r_s is the solution of (4.5.6).

We obtain:

$$|h_s(z)| \leq r_s + s\alpha(r_s) = 1, \qquad z \in \Delta.$$

Since $h_s(w) = J_s(w) \in \Delta$ it follows by the maximum principle that h_s maps Δ into itself. Therefore by Corollary 1.1.1

$$\frac{|h'_s(w)|}{1 - |h_s(w)|^2} \leq \frac{1}{1 - |w|^2}$$

for all $z \in \Delta$.

But

$$h'_s(z) = \left(1 + s\,\frac{\alpha(r_n)}{2}\right) [J_s(z)]'.$$

Hence

$$\frac{|[J_s(w)]'|}{1 - |J_s(w)|^2} \leq \frac{1}{1 + s\frac{\alpha(r_s)}{2}} \cdot \frac{1}{1 - |w|^2}.$$

In other words, the mapping J_s is a strict contraction with respect to the infinitesimal Poincaré metric $d\rho_w = \dfrac{|dw|}{1 - |w|^2}$, $w \in \Delta$, defined in Section 2.2.

Integrating the above inequality we obtain the same conclusion for the hyperbolic metric ρ:

$$\rho\left(J_s(z), J_s(w)\right) \leq \frac{1}{1 + s\frac{\alpha(r_s)}{2}}\rho(z, w). \tag{4.5.7}$$

Since $\alpha(r)$ is bounded on the interval $[0, 1]$ it follows by (4.5.6) that $r_s \to 1$ as $s \to 0$ and $\alpha(r_s) \to \alpha > 0$ because of its continuity. Therefore there exist some positive δ and ϵ such that for all $s \in (0, \delta)$ we have $\alpha(r_s) \geq \epsilon > 0$. Then for such s we obtain the inequality

$$\rho\left(J_s(z), J_s(w)\right) \leq \frac{1}{1 + s\epsilon}\,\rho(z, w),$$

which means that f is strongly ρ-monotone. Now setting $s = t/n$ in (4.5.7) and using the exponential formula:

$$F_t(z) = \lim_{n \to \infty} J_{t/n}^{(n)}(z)$$

we obtain by induction:

$$\begin{aligned}
\rho(F_t(z), F_t(w)) &\leq \lim_{h \to \infty} \frac{1}{\left(1 + \frac{t}{n} \cdot \frac{\alpha(r_n)}{2}\right)^n}\,\rho(z, w) \\
&= \exp\left\{-\frac{\alpha t}{2}\right\}\rho(z, w).
\end{aligned}$$

The proposition is proved. $\square$

Example 1. Let $f \in \text{Hol}(\Delta, \mathbb{C})$ be defined by

$$f(z) = a - \bar{a}z^2 + bz\frac{1-cz}{1+cz},$$

where $a, b \in \mathbb{C}$, $\text{Re}\,b > 0$, and $0 \leq c < 1$. If we introduce the function

$$\alpha(s) = -|a|(1-s^2) + (\text{Re}\,b)s\frac{1-cs}{1+cs},$$

then we obtain

$$\text{Re}\,g(z)\bar{z} \geq \alpha(|z|)|z|$$

and

$$\alpha(1) = \text{Re}\,b\frac{1-c}{1+c} > 0.$$

Hence $f(z)$ is a strongly semi-complete vector field on Δ.

Remark 4.5.1 Note that if $f \in \text{Hol}(\Delta, \mathbb{C})$ is known to be a semi-complete vector field on Δ, then condition (4.5.2) can be replaced by a slightly more general condition, namely,

$$\alpha(l) > 0 \quad \text{for some} \quad l \in (0, 1],$$

which still ensures f to be strongly semi-complete. The above arguments can be employed in the disk Δ_l. This note leads us to the following simple sufficient condition yielding the existence of an interior null point of a semi-complete vector field and its attractiveness.

Corollary 4.5.1 (cf., EM-RS-SD-2000c) *Let $f \in \mathcal{G}\,\text{Hol}(\Delta)$ be such that:*

$$\text{Re}\,f'(0) > 4|f(0)|. \tag{4.5.8}$$

Then f has a unique null point $\tau \in \Delta$ and the semigroup $S_f = \{F_t\}_{t\geq 0}$ converges to τ as t goes to infinity.

Proof. Consider the function:

$$\alpha(r) = -a(1-r^2) + br\frac{1-r}{1+r},$$

where $a = |f(0)|$, $b = \text{Re}\,f'(0)$. It follows by Proposition 3.5.3 that

$$\text{Re}\,f(z)\bar{z} \geq \alpha(|z|)|z|, \quad z \in \Delta. \tag{4.5.9}$$

Then we have under condition (4.5.8), that $\alpha(0) = -a < 0$, $\alpha(1) = 0$, and $\alpha'(1) < 0$. Hence, there is $l \in (0, 1)$, such that $\alpha(l) > 0$. $\square$

Remark 4.5.2 Note that conditions (4.5.8) and (4.5.9) imply that for some $\delta > 0$ each disk Δ_r with $r \in (1-\delta, r]$ is invariant for the semigroup $\{F_t\}$ generated by f. However these conditions are not sufficient to ensure the validity of the strong ρ-monotonicity of f.

Remark 4.5.3 Observe also that the above Proposition 4.5.1, Lemma 3.6.1, and Proposition 3.6.2 imply that the property of $f \in \mathrm{Hol}(\Delta, \mathbb{C})$ being strongly ρ-monotone is equivalent to the **strong flow invariance condition**

$$\mathrm{Re}\, f(z)\bar{z} \geq \eta > 0$$

for all z in the annulus $\{1 - \delta < |z| < 1\}$ for some $\delta > 0$.

In terms of the Poincaré metric this property is equivalent to the global uniform exponential convergence (see formula (4.5.3)) of the semigroup $\{F_t\}_{t \geq 0}$ generated by f on all of Δ, whilst the property of f to be strongly semi-complete is equivalent to local uniform exponential convergence on each compact subset of Δ (see Proposition 4.4.3).

4.6 Continuous version of the Julia–Wolff–Carathéodory Theorem

Let now $f \in \mathrm{Hol}(\Delta)$ be a null point free mapping. We already know that in this case the semigroup $S_f = \{F_t\}_{t > 0}$ generated by f is convergent (see Section 4.3). Moreover, there is a point $\tau \in \partial\Delta$ (a sink point of S_f), such that for each $z \in \Delta$, $\lim\limits_{t \to \infty} F_t(z) = \tau$. However, the study of rates of convergence in this case is more complicated than in the case when f has a null point in Δ. Indeed, as we saw above, in the latter case the asymptotic behavior of the semigroup generated by f is completely determined by the value of $f'(\tau)$. Namely, $\{F_t\}_{t > 0}$ is convergent if and only if $\mathrm{Re}\, f'(\tau) > 0$. Therefore it is natural that the rates of convergence obtained in Section 4.4 are connected with this value.

If $f \in \mathcal{G}\,\mathrm{Hol}(\Delta)$ is null point free, i.e., $S_f = \{F_t\}_{t \geq 0}$ has a sink point $\tau \in \partial\Delta$, one cannot use the same approach, since $f'(\tau)$ is not defined in general. In addition, the above approach of using a Möbius transformation is also unfeasible in this situation. Therefore we need another method to study the asymptotic behavior of the semigroup generated by a null point free holomorphic function.

A complete characterization of convergence of a semigroup to its sink point can be done by using the so called nontangential derivative of its generator in the spirit of the Julia–Wolff–Carathéodory Theorem (Proposition 1.4.2). Here we establish a continuous version of this result (Proposition 4.6.2) which will provide another proof of the classical one. Note in passing that if $F \in \mathrm{Hol}(\Delta)$ is a holomorphic self-mapping of Δ, then $f = I - F$ is semi-complete.

We first prove some auxiliary assertions.

Lemma 4.6.1 *Let* $f \in \text{Hol}(\Delta, \mathbb{C})$ *and let* $e \in \partial\Delta$. *If* $\beta := \angle \lim_{z \to e} \dfrac{f(z)}{z - e}$ *exists (finitely) then:*

(i) $\angle \lim_{z \to e} f(z) = 0$;

(ii) *the angular limit* $\angle \lim_{z \to e} f'(z)$ *also exists and equals to* β.

Proof. (i) is trivial. To prove (ii) let us assume that $e = 1$ and present f in the form

$$f(z) = \beta(z - 1) + h(z). \tag{4.6.1}$$

Then we have

$$\angle \lim_{z \to 1} h(z) = 0$$

and

$$\angle \lim_{z \to 1} \frac{h(z)}{z - 1} = 0. \tag{4.6.2}$$

We wish to show that

$$\angle \lim_{z \to 1} h'(z) = 0. \tag{4.6.3}$$

To this end take two sectors S and $\widetilde{S}$ in Δ both with vertex at $e = 1$, so that $S \subset \widetilde{S}$. For $z \in S$ we denote by $\Gamma(z)$ the circle with its center at z, such that $\Gamma(z)$ is tangent to the boundary of $\widetilde{S}$.

By θ we denote the angle between segments $[0, 1]$ and $[z, 1]$ and let $2\theta_s$ and $2\theta_{\widetilde{S}}$ be the angles of the sectors S and $\widetilde{S}$, respectively, at vertex $e = 1$ (see Fig. 4.2). Then

$$\sin(\theta_{\widetilde{S}} - \theta_S) \leq \sin(\theta_{\widetilde{S}} - \theta) = \frac{r(z)}{|1 - z|}, \tag{4.6.4}$$

where $r(z)$ is the radius of the circle $\Gamma(z)$. Now observe that when z converges to 1 in the sector S, all points $w \in \partial\Gamma(z)$ converge to 1 in the sector $\widetilde{S}$. Therefore using (4.6.2) we obtain that for each $\varepsilon > 0$ there is a point z close to 1 such that

$$\frac{|h(w)|}{|w - 1|} < \varepsilon, \quad w \in \Gamma(z).$$

It then follows by the Cauchy formula and (4.6.4) that

$$
\begin{aligned}
h'(z) &= \left| \frac{1}{2\pi i} \int_{\Gamma(z)} \frac{h(z)}{(w - z)^2} \, dw \right| \leq \frac{\varepsilon}{2\pi} \int_{\Gamma(z)} \frac{|w - 1|}{|w - z|^2} \, |dw| \\
&\leq \frac{\varepsilon}{2\pi} \max_{w \in \Gamma(z)} |w - 1| \int_{\Gamma(z)} \frac{|dw|}{|w - z|^2} = \frac{\varepsilon}{r(z)} \max_{w \in \Gamma(z)} |w - 1| \\
&= \frac{\varepsilon}{r(z)} \left(r(z) + |1 - z| \right) = \varepsilon \left(1 + \frac{|1 - z|}{r(z)} \right) \\
&\leq \varepsilon \left(1 + \frac{1}{\sin(\theta_{\widetilde{S}} - \theta_S)} \right).
\end{aligned}
$$

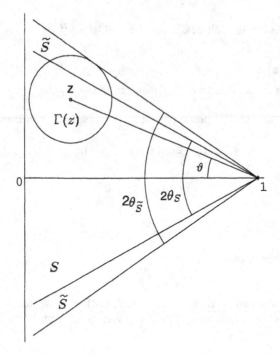

Figure 4.2: The circle $\Gamma(z)$ and the sectors S and $\widetilde{S}$.

The latter inequality concludes the proof of the Lemma. $\square$

The value $\beta := \angle \lim\limits_{z \to e} \dfrac{f(z)}{z - e}$ is called the angular derivative of f at $e \in \partial \Delta$.

We say that a function $f \in \mathrm{Hol}(\Delta, \mathbb{C})$ has the radial derivative at $e \in \partial \Delta$, if the limits

$$a := \lim_{r \to 1^-} f(re)$$

and

$$\lim_{r \to 1^-} \frac{f(re) - a}{(r - 1)e}$$

exist. We will denote it by $\uparrow f'(e)$.

It is clear that if f has the angular derivative at a point $e \in \partial \Delta$ then it has a radial derivative at this point and

$$\uparrow f'(e) = \angle f'(e).$$

We will show that for a null point free semi-complete vector field a somewhat converse statement is also true at the sink point of the flow generated by f. First we establish the following assertion.

Lemma 4.6.2 *Let f be a semi-complete vector field on Δ. Suppose that for a* *point $e \in \partial\Delta$ there exists the radial limit derivative*

$$\beta = \lim_{r \to 1^-} \frac{f(re)}{(r-1)e},$$

such that

$$\operatorname{Re}\beta \geq 0. \qquad (4.6.5)$$

Then f has no null point in Δ. Moreover, the point e must be the sink point of the semigroup generated by f.

Proof. Without loss of generality let as assume that $e = 1$, and write $f(z)$ by the Berkson–Porta representation

$$f(z) = (z - \tau)(1 - z\bar{\tau})p(z), \qquad (4.6.6)$$

where $\operatorname{Re}p(z) \geq 0$ everywhere, and τ is a point of $\overline{\Delta}$.

We wish to show that τ in (4.6.6) is equal to 1. First we assume that

$$\operatorname{Re}\beta > 0 \qquad (4.6.7)$$

and conversely suppose that $\tau \neq 1$. Then we have by (4.6.6) and (4.6.7)

$$\operatorname{Re}\beta = \operatorname{Re}\left(\lim_{r \to 1^-} \frac{f(r)}{r-1}\right) = \operatorname{Re}\left[\lim_{r \to 1^-}(r-\tau)(1-r\bar{\tau})\frac{p(r)}{r-1}\right]$$

$$= |1 - \bar{\tau}|^2 \operatorname{Re}\left[\lim_{r \to 1^-}\frac{\operatorname{Re}p(r)}{r-1}\right] > 0.$$

On the other hand for all $r \in (0,1)$

$$\operatorname{Re}\frac{p(r)}{r-1} = \frac{1}{r-1}\operatorname{Re}p(r) \leq 0,$$

which is a contradiction.

To complete our proof for the general case we consider the mapping $f_\varepsilon : \Delta \mapsto \mathbb{C}$ defined by

$$f_\varepsilon(z) = f(z) + \varepsilon(z^2 - 1) \qquad (4.6.8)$$

with $\varepsilon > 0$. Since $\varepsilon(z^2 - 1)$ is a complete vector field, the vector field f_ε is semicomplete and

$$\operatorname{Re}[\uparrow f_\varepsilon'(1)] = \operatorname{Re}\beta + 2\varepsilon > 0. \qquad (4.6.9)$$

Therefore by the previous step, for each $\varepsilon > 0$ the mapping f_ε has no null points in Δ. Now, if τ in (4.6.6) belongs to Δ, that is, f has an interior null point, then by the Rouchet theorem it follows that one can choose a small enough ε such that f_ε has an interior null point τ_ε close to τ. Once again a contradiction. Moreover, (4.6.9) and the step proved above show that for each $\varepsilon > 0$ the point 1 must be a sink point of f_ε. Thus there are functions $p_\varepsilon \in \operatorname{Hol}(\Delta, \mathbb{C})$ with $\operatorname{Re}p_\varepsilon \geq 0$, $z \in \Delta$, such that

$$f_\varepsilon(z) = -(1 - z)^2 p_\varepsilon(z).$$

Comparing the latter formula with (4.6.8) we obtain

$$p_\varepsilon(z) = \frac{(z-\tau)(1-z\bar\tau)p(z)}{-(1-z)^2} + \varepsilon\frac{z+1}{1-z}.$$

Since $\operatorname{Re} p_\varepsilon \geq 0$ for all $\varepsilon > 0$ and $p_\varepsilon(z)$ converges to $p_0(z)$

$$p_0(z) = \frac{(z-\tau)(1-z\bar\tau)p(z)}{-(1-z)^2},$$

uniformly on each compact subset of Δ as ε goes to 0, we obtain that

$$\operatorname{Re} p_0 \geq 0.$$

But

$$-(1-z)^2 p_0(z) = (z-\tau)(1-z\bar\tau)p(z) = f(z),$$

hence contradicting the uniqueness of the Berkson–Porta representation.
Thus $p_0(z) = p(z)$ and $\tau = 1$. $\square$

Proposition 4.6.1 *If for a point $e \in \partial\Delta$ the radial limits*

$$\uparrow \lim_{z\to e} f'(z) = \beta \quad and \quad \uparrow \lim_{z\to e} f(z) = 0$$

exist with $\operatorname{Re}\beta \geq 0$, then

$$0 \leq \beta = \angle f'(e)$$

and $e = \tau$ is a sink point of the flow generated by f.

Proof. Indeed, since $|f'(re)| < M < \infty$ we obtain

$$|f(re)| = \left| \int_r^1 f'(te)\, dt \right| \leq M(1-r).$$

Hence $\displaystyle\lim_{r\to 1^-}\inf \frac{f(re)}{(r-1)e}$ exists. Let $e = 1$ and $u(z) = \operatorname{Re} f(z)$. Then we obtain

$$u(r) = -\int_r^1 u'(t)dt = u'(\eta)(r-1),$$

where $\eta \in [r,1)$. Hence

$$\lim_{r\to 1^-} \frac{u(r)}{(r-1)} = \operatorname{Re}\beta \geq 0. \ \square$$

To establish the converse assertion to Lemma 4.6.2 we will use the Riesz–Herglotz integral representation of functions of the class $\mathcal{P} = \{p \in \operatorname{Hol}(\Delta, \mathbb{C}) : \operatorname{Re} p(z) \geq 0\}$ (see Section 0.2):

$$p(z) = \int_{\partial\Delta} \frac{1+z\bar\zeta}{1-z\bar\zeta}\, d\mu_p(\zeta) + i\operatorname{Im} p(0), \tag{4.6.10}$$

where $\mu_p : \partial\Delta \mapsto \mathbb{R}$ (the measure characteristic function for $p \in \mathcal{P}$) is a positive increasing finite function μ_p on the unit circle $\partial\Delta$, such that

$$\int_{\partial\Delta} d\mu_p(\zeta) = \operatorname{Re} p(0).$$

Lemma 4.6.3 *Let* $f \in \operatorname{Hol}(\Delta, \mathbb{C})$ *be a semi-complete vector field on* Δ *with no null point in* Δ. *Then there is a point* $\tau \in \partial\Delta$, *such that* f *has an angular derivative at* τ. *Moreover, if* τ *is a sink point of the flow generated by* f *then* $\angle f'(\tau)$ *exists and it is a positive real number which is equal to* $2\mu_p(\tau)$, *where*

$$p(z) = \frac{f(z)}{(z-\tau)(1-z\bar{\tau})}.$$

Proof. Of course, it is enough to prove only the second assertion of the Lemma. Let us again present f by the Berkson–Porta formula

$$f(z) = (z-\tau)(1-z\bar{\tau})p(z),$$

where $\operatorname{Re} p(z) \geq 0$, $z \in \Delta$. Then by (4.6.10) one can write

$$\frac{f(z)}{(z-\tau)} = \int_{\partial\Delta} (1-z\bar{\tau})\frac{1+z\bar{\zeta}}{1-z\bar{\zeta}} d\mu_p(\zeta) - (1-z\bar{\tau})i\operatorname{Im}\tau f(0). \tag{4.6.11}$$

Let $z_n \in \Delta$ be a sequence of points which converges to τ nontangentially. Then if we write z_n in the form

$$z_n = \tau(x_n + iy_n),$$

we can find $0 < K < \infty$, such that

$$|y_n| \leq K(1-x_n) \tag{4.6.12}$$

(see Fig. 4.3).

Further, consider the sequence of functions $g_n : \partial\Delta \mapsto \mathbb{C}$ defined by

$$g_n := g_n(\zeta) = \frac{(1-\bar{\tau}z_n)(1+z_n\bar{\zeta})}{1-z_n\bar{\zeta}}. \tag{4.6.13}$$

To estimate g_n on the unit circle we calculate

$$(1-\bar{\tau}z_n)(1+z_n\bar{\zeta}) = w_n(1-z_n\bar{\zeta})$$

or

$$z_n\bar{\zeta}(1-\bar{\tau}z_n + g_n) = g_n - (1-\bar{\tau}z_n).$$

This implies that

$$|z_n|^2|1-\bar{\tau}z_n + g_n|^2 = |g_n - (1-\bar{\tau}z_n)|^2$$

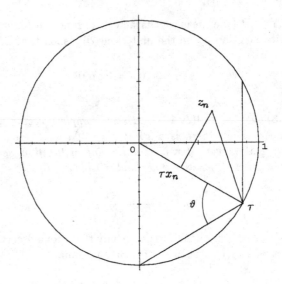

Figure 4.3: The nontangential convergence to τ.

or

$$|g_n|^2(1 - |z_n|^2) - 2\operatorname{Re}[g_n(1 - \bar{\tau}z_n)](1 + |z_n|^2) = |1 - \bar{\tau}z_n|^2(|z_n|^2 - 1).$$

Consequently,

$$\left|g_n - \frac{(1 - \bar{\tau}z_n)(1 + |z_n|^2)}{1 - |z_n|^2}\right|^2 = \frac{|1 - \bar{\tau}z_n|^2(1 + |z_n|^2)^2}{(1 - |z_n|^2)^2} - |1 - \bar{\tau}z_n|^2$$

$$= 4|z_n|^2\frac{|1 - \bar{\tau}z_n|^2}{(1 - |z_n|^2)^2}.$$

Finally,

$$\left|g_n - \frac{(1 - \bar{\tau}z_n)(1 + |z_n|^2)}{1 - |z_n|^2}\right| = 2|z_n|\frac{|1 - \bar{\tau}z_n|}{1 - |z_n|^2},$$

and we obtain

$$|w_n(\zeta)| \le 4\frac{|1 - \bar{\tau}z_n|}{1 - |z_n|^2} = 4\frac{\sqrt{(1 - x_n)^2 + y_n^2}}{(1 - x_n)^2 y_n^2}.$$

Taking (4.6.12) into account we obtain

$$|w_n(\varphi)| \le 4\frac{\sqrt{1 + \left(\frac{y_n}{(1-x_n)}\right)^2}}{1 + x_n - |y_n|\left|\frac{y_n}{1-x_n}\right|} = 4\frac{\sqrt{1 + K^2}}{1 + x_n - |y_n|K}.$$

Since for n large enough

$$1 + x_n - |y_n|K > \frac{1}{2},$$

we have for such n

$$|g_n(\zeta)| \leq 8\sqrt{1 + K^2}.$$

In addition, it follows by (4.6.13) that

$$g_n(\zeta) \to 0, \quad \zeta \neq \tau.$$

Then by the Lebesgue Theorem (see, for example, [125]) and (4.6.11) we obtain

$$\lim_{z_n \to \tau} \frac{f(z_n)}{z_n - \tau} = \int_{\partial \Delta} \frac{(1 - \bar{\tau} z_n)(1 + z_n \bar{\zeta})}{1 - z_n \bar{\zeta}} \, d\mu_p(\zeta) - i \operatorname{Im} \tau f(0) \cdot \lim_{z_n \to \tau} (1 - z_n \bar{\tau})$$

$$= 2\mu_p(\tau),$$

and the Lemma is proved. $\square$

Our next goal is to establish relations between angular derivatives of semi-complete vector fields and the asymptotic behavior of their generated semigroups.

Lemma 4.6.4 *Let $F \in \mathrm{Hol}(\Delta, \Delta)$ be a holomorphic self-mapping of Δ with no null point in Δ, and let $\tau \in \partial\Delta$ be its sink point. If z_n converges to τ nontangentially, then so does the sequence $F(z_n)$.*

Proof. It follows by the Julia–Wolff–Carathéodory Theorem (Proposition 1.4.2) that

$$0 \leq \alpha = \angle F'(\tau) \leq 1$$

and

$$\varphi_\tau(F(z)) \leq \alpha \varphi_\tau(z), \tag{4.6.14}$$

where

$$\varphi_\tau(z) = \frac{|1 - z\bar{\tau}|^2}{1 - |z|^2} = \frac{|z - \tau|^2}{1 - |z|^2}.$$

If $z \in \Gamma(\tau, k) = \{z \in \Delta : |z - \tau| < k(1 - |z|)\}$ for some $k > 1$ (see Definition 1.2.1) then (4.6.14) implies

$$\frac{|F(z) - \tau|}{1 - |F(z)|} \leq \alpha \frac{|z - \tau|}{1 - |z|} \cdot \frac{|z - \tau|}{|F(z) - \tau|} \cdot \frac{1 + |F(z)|}{1 + |z|}. \tag{4.6.15}$$

In turn, The Julia–Carathéodory Theorem states that

$$\angle \lim_{z_n \to \tau} \frac{F(z) - \tau}{z - \tau} = \alpha > 0.$$

Therefore the third factor in (4.6.15) is bounded. Thus we can find $k_1 \geq k$, such that $F(z) \in \Gamma(\tau, k_1)$ and we are done. $\square$

Finally we are able to formulate our main assertion of this section which is a dynamical analog of the Julia–Wolff–Carathéodory Theorem.

Proposition 4.6.2 (see [42]) *Let $f \in \mathcal{G}\,\mathrm{Hol}(\Delta, \mathbb{C})$ be a semi-complete vector field, and let $S = \{F_t\}_{t \geq 0}$ be the semigroup generated by f. The following are equivalent:*

(i)   *f has no null point in Δ;*
(ii)   *f admits the representation*

$$f(z) = -\bar{\tau}(z - \tau)^2 p(z)$$

for some $\tau \in \partial\Delta$ and $\mathrm{Re}\, p(z) \geq 0$ everywhere;
(iii) *there is a point $\tau \in \partial\Delta$, such that*

$$\uparrow f'(\tau) = \beta$$

exists and $\mathrm{Re}\,\beta \geq 0$;
(iv) *there is a point $\tau \in \partial\Delta$, such that*

$$\angle f'(\tau) = \beta$$

exists and $\mathrm{Re}\,\beta \geq 0$;
(v) *there is a point $\tau \in \partial\Delta$, such that the following limits exist*

$$\angle \lim_{z \to \tau} f'(z) = \beta \quad and \quad \angle \lim_{z \to \tau} f(z) = 0$$

with $\mathrm{Re}\,\beta \geq 0$;
(vi) *there are a point $\tau \in \partial\Delta$ and a real positive number γ, such that*

$$\frac{|F_t(z) - \tau|^2}{1 - |F_t(z)|^2} \leq e^{-t\gamma} \frac{|z - \tau|^2}{1 - |z|^2}.$$

Moreover,

(a) the points $\tau \in \partial\Delta$ in (ii)–(vi) and the numbers β in (iii)–(v) are the same;
(b) β is, in fact, a nonnegative real number which is the maximum of all $\gamma \geq 0$ which satisfy (vi).

Proof. Equivalence of (i)–(iv) has been proved in Lemmata 4.6.2–4.6.3, while equivalence (iv) and (v) follows from Lemma 4.6.1 and Proposition 4.6.1. Thus it is enough to show that (v)$\Rightarrow$(vi) and (vi) implies one of the conditions (i)–(v). Suppose that (v) holds. We already know that f has no null point in Δ and $e = \tau$ is the sink point of its generated semigroup. In addition, if $J_r \in \mathrm{Hol}(\Delta)$, $r \geq 0$, is a resolvent of f:

$$J_r(z) + rf(J_r(z)) = z, \quad z \in \Delta, \tag{4.6.16}$$

then τ is also the sink point for $J_r, r > 0$. It follows by Lemma 4.6.4 that if z converges nontangentially to τ, then so does $J_r(z)$ for each $r \geq 0$.
Hence

$$\angle \lim_{z \to \tau} f'(J_r(z)) = \beta. \tag{4.6.17}$$

On the other hand, it follows by Julia's Lemma that for each $r > 0$ there is a number α_r, $0 < \alpha_r \leq 1$, such that

$$\angle \lim_{z \to \tau} J_r'(z) = \alpha_r \tag{4.6.18}$$

and

$$\frac{|J_r(z) - \tau|^2}{1 - |J_r(z)|^2} \leq \alpha_r \frac{|z - \tau|^2}{1 - |z|^2}. \tag{4.6.19}$$

At the same time, differentiating (4.6.16) we obtain .

$$J_r'(z) + rf'(J_r(z)) \cdot J_r'(z) = 1,$$

or

$$J_r'(z) = \frac{1}{1 + rf'(J_r(z))}.$$

Using (4.6.17) and (4.6.18) we obtain

$$\alpha_r = \frac{1}{1 + r\beta}. \tag{4.6.20}$$

Substituting (4.6.20) into (4.6.19) and applying the exponential formula we obtain

$$\begin{aligned}
\varphi_\tau(F_t(z)) &= \lim_{n \to \infty} \varphi_\tau(J_{t/n}^{(n)}(z)) \\
&\leq \lim_{n \to \infty} \frac{1}{\left(1 + \frac{t}{n}\beta\right)^n} \varphi_\tau(z) = \exp\{-t\beta\}\varphi_\tau(z),
\end{aligned} \tag{4.6.21}$$

where

$$\varphi_\tau(z) = \frac{|z - \tau|^2}{1 - |z|^2}. \tag{4.6.22}$$

Thus the implication (v)$\Rightarrow$(vi) is proved. Obviously (vi)$\Rightarrow$(i), because in this case the semigroup $\{F_t\}$ has no stationary point in Δ. Then it remains to prove assertion (b). In other words, we wish to show that if (vi) holds with some $\gamma \geq 0$ then $\gamma \leq \beta$. Indeed, consider the real valued function

$$\psi(t, z) = \varphi_\tau(F_t(z)), \tag{4.6.23}$$

where φ_τ is defined in (4.6.22). Since by (vi) $\psi(t, z) \leq e^{-\gamma t}\psi(0, z)$ we have

$$\begin{aligned}
\frac{\partial \psi}{\partial t}\Big|_{t=0^+} &= \lim_{t \to 0^+} \frac{\psi(t, z) - \psi(0, z)}{t} \\
&\leq \lim_{t \to 0^+} \frac{\psi(0, z)(e^{-\gamma t} - 1)}{t} = -\gamma\psi(0, z).
\end{aligned} \tag{4.6.24}$$

On the other hand, differentiating (4.6.23) directly and using

$$\frac{\partial F_t(z)}{\partial t}\Big|_{t=0^+} = -f(z)$$

we obtain

$$\frac{\partial \psi}{\partial t}\Big|_{t=0^+} = -2\psi(0, z) \cdot \operatorname{Re} f(z)z^*, \tag{4.6.25}$$

where

$$z^* = \frac{\bar{z}}{1 - |z|^2} - \frac{\bar{\tau}}{1 - z\bar{\tau}}.$$

Comparing (4.6.24) with (4.6.25) we obtain

$$\gamma \leq 2\operatorname{Re} f(z) z^*. \tag{4.6.26}$$

Let us again suppose for simplicity that $\tau = 1$. Then we may rewrite the right hand side of (4.6.26) in the form:

$$2\operatorname{Re} f(z)z^* = 2\operatorname{Re} f(z)\left(\frac{\bar{z}}{1 - |z|^2} - \frac{1}{1 - z}\right) = 2\operatorname{Re} \frac{f(z)(\bar{z} - 1)}{(1 - z)(1 - |z|^2)}.$$

Setting here $z = r$ and letting $r \to 1^-$ we obtain by (4.6.26):

$$\gamma \leq 2 \lim_{r \to 1^-} \operatorname{Re} \frac{f(r)}{(r - 1)(r + 1)} = \lim_{r \to 1^-} \frac{f(r)}{(r - 1)} = \beta.$$

The proof is completed. □

This result will play a crucial role in our study of the spirallike and starlike functions with respect to a boundary point (see Chapter 5).

At the end of this section we consider some examples and a consequence of the above proposition which is an extension of the Julia–Wolff–Carathéodory Theorem.

Example 1. Consider a semi-complete vector field $f \in \operatorname{Hol}(\Delta, \mathbb{C})$ defined as follows

$$f(z) = (1 - z)^2 \cdot \frac{z^n + 1}{z^n - 1}.$$

Computations show that $f'(1) = 2/n > 0$, while $\operatorname{Re} f'(z_k) < 0$, where $z_k^n = -1$, $k = 1, 2, \ldots, n$. Hence if $\{F_t\}_{t \geq 0}$ is the semigroup generated by f, then the point $\tau = 1$ is an attractive point of this semigroup and the following rate of exponential convergence holds

$$\frac{|F_t(z) - 1|^2}{1 - |F_t(z)|^2} \leq e^{-\frac{2t}{n}} \frac{|z - 1|^2}{1 - |z|^2}$$

for all $z \in \Delta$ and $t \geq 0$.

As we mentioned above, if F is a self-mapping of Δ then the function $f(z) = z - F(z)$ defines a semi-complete vector field on Δ. As a matter of fact, this fact holds even if $F \in \operatorname{Hol}(\Delta, \mathbb{C})$ is not necessarily a self-mapping of Δ, but satisfies the following one-sided estimate:

$$\lim_{r \to 1^-} F(rz)\bar{z} \leq 1 \quad \text{for all } z \in \partial\Delta. \tag{4.6.27}$$

Thus we have the following version of the Julia–Wolff–Carathéodory Theorem:

Corollary 4.6.1 *Let* $F \in \mathrm{Hol}(\Delta, \Delta)$ *satisfy (4.6.27) and* $z \in \partial\Delta$. *Then the following statements are equivalent:*

(i) *F has no fixed point in* Δ;

(ii) *for some* $\omega \in \partial\Delta$ *there exists the angular limit*

$$\angle \lim_{z \to \omega} \frac{F(z) - \omega}{z - \omega} := \alpha$$

with $\mathrm{Re}\,\alpha \le 1$;

(iii) *F admits the representation* $F(z) = z + (z - \omega)^2 \bar{\omega} p(z)$ *for some* $\omega \in \partial\Delta$ *and* $p \in \mathrm{Hol}(\Delta, \mathbb{C})$ *with* $\mathrm{Re}\,p(z) \ge 0$.

Moreover, α *in (ii) is actually a real number and the boundary points* ω *in (ii) and (iii) are the same.*

4.7 Lower bounds for ρ-monotone functions

We have seen in previous sections that the asymptotic behavior of semigroups generated by holomorphic functions can be described in terms of their derivatives.

If $f \in \mathcal{G}\,\mathrm{Hol}(\Delta)$ and $\tau \in \mathrm{Null}(f)$ then τ is (globally) attractive if and only if $f'(\tau)$, the derivative of f at τ, lies strictly in the right half-plane.

Moreover, $f \in \mathcal{G}\,\mathrm{Hol}(\Delta)$ has no null point in Δ if and only if for some $\tau \in \partial\Delta$ the angular derivative

$$\angle f'(\tau) = \beta$$

exists (finitely) with $\beta \ge 0$.

In addition, if $S = \{F_t\}_{t \ge 0}$ is the flow generated by f then

$$\frac{|F_t(z) - \tau|^2}{1 - |F_t(z)|^2} \le \exp\left(-t\beta\right) \frac{|z - \tau|^2}{1 - |z|^2},$$

and the point $\tau \in \partial\Delta$ is a (globally) attractive sink point of S (even if $\beta = 0$).

However, if $\angle f'(\tau) = \beta = 0$ the latter formula does not help to establish a rate of convergence of the semigroup to its sink point.

Note also that if $f \in \mathcal{G}N_\rho(\Delta)$ is not holomorphic the characteristics of the derivatives are not relevant.

Actually, we will show that for $f \in \mathcal{G}\,\mathrm{Hol}(\Delta)$ the number $\beta = \angle f'(\tau)$ is equal to

$$\inf_{z \in \Delta} 2\,\mathrm{Re}\,f(z)\overline{z^*}, \tag{4.7.1}$$

where

$$z^* = \frac{z}{1 - |z|^2} - \frac{\tau}{1 - \bar{z}\tau}.$$

It turns out that even if $f \in \mathcal{G}N_\rho(\Delta)$ is not holomorphic, expression (4.7.1) can serve as a characterization of the asymptotic behavior of flows of ρ-nonexpansive mappings in Δ both in the cases of an interior stationary point and a boundary sink point.

In this section we will mostly follow the material of [39].

For a fixed $\tau \in \bar{\Delta}$, the closure of Δ, and an arbitrary $z \in \Delta$, we define a non-Euclidean 'distance' between z to τ by the formula:

$$d_\tau(z) = \frac{|1 - z\bar{\tau}|^2}{1 - |z|^2} \left(1 - \sigma(z, \tau)\right), \tag{4.7.2}$$

where

$$\sigma(z, \tau) = \frac{\left(1 - |z|^2\right)\left(1 - |\tau|^2\right)}{|1 - z\bar{\tau}|^2}$$

(see Section 2.3).

Exercise 1. Show that the sets

$$E(\tau, s) = \left\{z \in \Delta : d_\tau(z) < s\right\}, \quad s > 0,$$

have the following geometric interpretation:

(a) If $\tau \in \Delta$, then these sets are exactly the ρ-balls

$$E(\tau, s) = B(\tau, r) = \left\{z \in \Delta : \rho(z, \tau) < r\right\}$$

centered at $\tau \in \Delta$ and of radius $r = \tanh^{-1} \sqrt{\dfrac{s}{s + 1 - |\tau|^2}}.$

(b) If $\tau \in \partial\Delta$, the boundary of Δ, then these sets

$$E(\tau, s) = D(\tau, s) = \left\{z \in \Delta : d_\tau(z) = \frac{|1 - z\bar{\tau}|^2}{1 - |z|^2} < s\right\}, \quad s > 0,$$

are horocycles in Δ which are internally tangent to the unit circle $\partial\Delta$ at τ.

Now for fixed $\tau \in \bar{\Delta}$ and $z \in \partial E(\tau, s) = \left\{z \in \Delta : d_\tau(z) = s\right\}$, $s > 0$, $z \neq \tau$, consider the nonzero vector

$$z^* = \frac{1}{1 - \sigma(z, \tau)} \left(\frac{1}{1 - |z|^2} z - \frac{1}{1 - z\bar{\tau}} \tau\right). \tag{4.7.3}$$

Exercise 2 ([6]). Show that z^* is a so called support functional of the smooth convex set $E(\tau, s)$ at the point $z \in \Delta$, $d_\tau(z) = s$, i.e., for all $w \in \overline{E(\tau, s)}$ the following inequality holds:

$$\mathrm{Re}(w\bar{z^*}) \leq \mathrm{Re}(z\bar{z^*}).$$

In order to classify the asymptotic behavior of a flow generated by $f \in \mathcal{G}N_\rho(\Delta)$ for a point $\tau \in \overline{\Delta}$ we consider two real nonnegative functions on $(0, \infty)$:

$$\omega_b(s) := \inf_{d_\tau(z) \leq s} 2 \operatorname{Re} f(z)\overline{z^*}, \quad s > 0, \qquad (4.7.4)$$

and

$$\omega^\sharp(s) := \inf_{d_\tau(z) = s} 2 \operatorname{Re} f(z)\overline{z^*}, \quad s > 0, \qquad (4.7.5)$$

where z^* is defined by (4.7.3).

If $\tau \in \overline{\Delta}$ is a stationary (or sink) point for the flow generated by $f \in \mathcal{G}N_\rho(\Delta)$, then it follows that $\operatorname{Re} f(z)\overline{z^*} \geq 0$ by ρ-monotonicity of f (see Section 3.4). Hence

$$\omega^\sharp(s) \geq \omega_b(s) \geq 0$$

and $\omega_b(s)$ is clearly decreasing on $(0, \infty)$.

Let $\mathcal{M}(0, \infty)$ denote the class of all positive functions ω on $(0, \infty)$ such that $\frac{1}{\omega}$ is Riemann integrable on each closed interval $[a, b] \subset (0, \infty)$ and

$$(*) \qquad \int\limits_{0+} \frac{ds}{\omega(s)s} \text{ is divergent.}$$

Note that for each $\omega \in \mathcal{M}(0, \infty)$ the function Ω defined by

$$\Omega(s) := \int\limits_s^{d_\tau(z)} \frac{d\lambda}{\omega(\lambda)\lambda} \qquad (4.7.6)$$

is a strictly decreasing positive function on $(0, d_\tau(z)]$ and maps this interval onto $[0, \infty)$. We denote its inverse function by $V : [0, \infty) \mapsto (0, d_\tau(z)]$.

Definition 4.7.1 *We will call a function $\omega \in \mathcal{M}(0, \infty)$ an appropriate lower bound for $f \in \mathcal{G}N_\rho(\Delta)$ if*

$$\omega(s) \leq \omega^\sharp(s) = \inf_{d_\tau(z) = s} 2 \operatorname{Re} f(z)\overline{z^*}, \quad s > 0.$$

Exercise 3. Show that if ω_b defined by (4.7.4) is not zero then it is an appropriate lower bound for $f \in \mathcal{G}N_\rho(\Delta)$.

Exercise 4. Let $p \in \operatorname{Hol}(\Delta, \mathbb{C})$ be such that $\operatorname{Re} p(z) \geq 0$, $z \in \Delta$ and

$$\angle \lim_{z \to 1} (1 - z)p(z) = \beta \geq 0.$$

Show that if $f \in \mathcal{G}\operatorname{Hol}(\Delta)$ is defined by the Berkson–Porta formula (see Section 3.6)

$$f(z) = -(1 - z)^2 p(z),$$

then for the sink point $\tau = 1$ the function ω_b is constant which is equal to β.

Proposition 4.7.1 *Let $f \in \mathcal{GN}_\rho(\Delta)$ be continuous and let $S = \{F_t\}_{t\geq 0}$ be the flow generated by f. Given a point $\tau \in \overline{\Delta}$ and a function $\omega \in \mathcal{M}(0,\infty)$, the following conditions are equivalent:*

(i) the function ω is an appropriate lower bound for f;

(ii) for any differentiable function W on $[0,\infty)$ such that $V(t) \leq W(t)$, $V(0) = W(0)$ and $V'(0) = W'(0)$,

$$d_\tau(F_t(z)) \leq W(t), \quad z \in \Delta, \ t \geq 0,$$

where $V = \Omega^{-1}$ and Ω is defined by (4.7.6).

In particular, $d_\tau(F_t(z)) \leq V(t)$; hence τ is a globally attractive stationary point for S.

Proof. Consider the function $\Psi : \mathbb{R}^+ \times \Delta \mapsto \mathbb{R}^+$ defined by

$$\Psi(t, z) = d_\tau(F_t(z)). \tag{4.7.7}$$

By direct calculations we have

$$\frac{\partial \Psi}{\partial t}\Big|_{t=0^+} = -2\Psi(0, z)\, \mathrm{Re}\, f(z)\overline{z^*}. \tag{4.7.8}$$

We first assume that condition (ii) holds. Since $\Psi(0, z) = d_\tau(z) = W(0)$, we obtain by (4.7.8) and (ii) that

$$
\begin{aligned}
2\Psi(0, z)\, \mathrm{Re}\, f(z)\overline{z^*} &= -\frac{\partial \Psi}{\partial t}\Big|_{t=0^+} \geq -\frac{d}{dt}[W(t)]_{t=0^+} \\
&= -\frac{d}{dt}[V(t)]_{t=0^+} = -\frac{1}{\Omega'(d_\tau(z))} \\
&= d_\tau(z)\omega(d_\tau(z)).
\end{aligned}
$$

Varying $z \in \partial E(\tau, s) = \{z \in \Delta : d_\tau(z) = s\}$ we see that the latter inequality immediately implies (i).

Conversely, let condition (i) hold. It follows by (4.7.7) and the semigroup property that for all $z \in \Delta$ and $s, t \geq 0$,

$$\Psi(s + t, z) = \Psi(s, F_t(z)).$$

Hence by (4.7.8) and the continuity of f, Ψ is differentiable at each $t \geq 0$ and we deduce from (i) and (4.7.8) that:

$$\frac{\partial \Psi(t, z)}{\partial t} \leq -\Psi(t, z)\omega^\sharp(\Psi(t, z)) \leq -\Psi(t, z)\omega(\Psi(t, z)).$$

Separating variables we obtain

$$\int_{d_\tau(F_t(z))}^{d_\tau(z)} \frac{d\Psi}{\omega(\Psi)\Psi} = \Omega(d_\tau F_t(z)) \geq t.$$

This is equivalent to condition (ii). Our assertion is proved. □

As we mentioned above, if $\omega_b(s) := \inf_{d_\tau(z)} \leq s\, 2\,\mathrm{Re}\, f(z)\overline{z^*}$ is positive then it may be used as an appropriate lower bound for $f \in \mathcal{G}N_\rho(\Delta)$ (see Example 1).

Example 1. Let $f : \Delta \mapsto \mathbb{C}$ be defined by

$$f(z) = -(1-z)^2 \frac{1+z^n}{1-z^n},$$

where n is a positive integer.

Since f is holomorphic on Δ it follows by the Berkson–Porta representation (see Section 3.6) that f generates a flow $S = \{F_t\}_{t\geq 0}$ of holomorphic self-mappings of Δ and $\tau = 1$ is a sink point of S.

If we now set

$$z^* = \frac{z}{1-|z|^2} - \frac{1}{1-\bar{z}}$$

then we obtain

$$\mathrm{Re}\, f(z)\overline{z^*} = \frac{|1-z|^2}{1-|z|^2}\, \mathrm{Re}\, \frac{1+z^n}{1-z^n} = d_1(z)\, \mathrm{Re}\, \frac{1+z^n}{1-z^n} > 0.$$

In addition, it can be shown (see Exercise 4 and Proposition 4.7.5 below) that

$$\omega(s) = \omega_b(s) = \inf_{d_1(z)\leq s} 2\,\mathrm{Re}\, f(z)\overline{z^*} = \text{constant} = \frac{2}{n}.$$

In this case

$$\Omega(s) = \frac{n}{2} \int_s^{d_1(z)} \frac{d\lambda}{\lambda} = -\frac{n}{2}\ln\frac{s}{d_1(z)}$$

and

$$V(t) = \Omega^{-1}(t) = \exp\left(-\frac{2}{n}t\right) d_1(z).$$

Thus we have an exponential rate of convergence of the flow S to the boundary point $\tau = 1$:

$$\frac{|F_t(z)-1|^2}{1-|F_t(z)|^2} \leq \exp\left\{-\frac{2}{n}t\right\} \frac{|1-z|^2}{1-|z|^2}.$$

Note also that although f has $n+1$ null points $\{a_k : k = 1,2,\ldots,n+1\}$ on the unit circle, only $a_1 = 1$ is an attractive point of $S = \{F_t\}_{t\geq 0}$. The reason is that $\mathrm{Re}\, f'(a_1) > 0$, while $\mathrm{Re}\, f'(a_k) < 0$, $k = 2,3,\ldots,n+1$ (see Figure 4.4 for $n = 3$).

Remark 4.7.1. However, examples show (see Example 2 below) that sometimes ω_b may be zero identically, while $\omega^\sharp$ itself belongs to the class $\mathcal{M}(0,\infty)$. Moreover, we will see below that for a semigroup of holomorphic mappings with a boundary sink point τ the function ω_b is always a constant which coincides with the angular

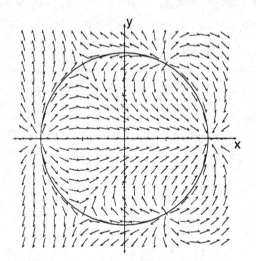

Figure 4.4: The asymptotic behavior of the flow generated by
$$f(z) = -(1-z)^2 \frac{1+z^3}{1-z^3}.$$

derivative of f at τ. Also observe that the same estimate as in Example 1 can be obtained by using Proposition 4.6.2. Nevertheless, even for holomorphic mappings Proposition 4.7.1 becomes an effective tool when the angular derivative $\angle f'(\tau) = 0$.

Example 2. Let $f : \Delta \mapsto \mathbb{C}$ be defined by

$$f(z) = -(1-z)^2 \frac{1+cz^n}{1-cz^n}$$

with $|c| < 1$.

Once again, if we define z^* as in Example 1 we have

$$\operatorname{Re} f(z)\overline{z^*} = \frac{|1-z|^2}{1-|z|^2} \operatorname{Re} \frac{1+cz^n}{1-cz^n} \geq d_1(z)\frac{1-|c|}{1+|c|} > 0.$$

In this case $\angle f'(1) = \omega_b(s) = 0$ for all $s \in (0, \infty)$ and we cannot use it as an appropriate lower bound. However, we can define $\omega(s) = as$, where $a = \dfrac{1-|c|}{1+|c|}$, and we find

$$\Omega(s) = \frac{1}{a} \int\limits_{s}^{d_1(z)} \frac{d\lambda}{\lambda^2} = \frac{1}{a}\left(\frac{1}{s} - \frac{1}{d_1(z)}\right).$$

Thus we obtain by Proposition 4.7.1 the following rate of nonexponential convergence:

$$\frac{|1 - F_t(z)|^2}{1 - |F_t(z)|^2} \leq \frac{1}{1 + atd_1(z)} \frac{|1 - z|^2}{1 - |z|^2}.$$

The reason is that $f \in \mathcal{G}\,\mathrm{Hol}(\Delta)$ has no null point inside Δ and $\omega_b(s)$ defined at the boundary sink point $\tau = 1$ is equal to zero.

Next we consider the case when $f \in \mathcal{G}N_\rho(\Delta)$ is not holomorphic on Δ. In this situation the convergence of the flow generated by f may be of nonexponential type even f has an interior null point.

Example 3. Let $z = x + iy \in \Delta$. Define $f : \Delta \mapsto \mathbb{C}$ by

$$f(z) = x^{7/3} + iy^{7/3}.$$

Since

$$\mathrm{Re}\, f(z)\bar{z} = x^{10/3} + y^{10/3} \geq 0,$$

f is ρ-monotone and the origin is the unique null point of f.

Then, setting $\tau = 0$, we have

$$d_0(z) = \frac{|z|^2}{1 - |z|^2}$$

and

$$
\begin{aligned}
\omega^\sharp(s) &= 2 \inf_{d_0(z)=s} \frac{1}{|z|^2(1 - |z|^2)}\, \mathrm{Re}\, f(z)\bar{z} \\
&= 2 \inf_{x^2+y^2=\frac{s}{s+1}} \frac{x^{10/3} + y^{10/3}}{(x^2 + y^2)(1 - x^2 - y^2)} \\
&= 2^{1/3} s^{2/3}(1 + s)^{1/3}.
\end{aligned}
$$

Since $\omega^\sharp(s) \in \mathcal{M}(0, \infty)$ we can set $\omega(s) = \omega^\sharp(s)$ and we have

$$\Omega(s) = \int_s^{d_0(z)} \frac{d\lambda}{\omega(\lambda)\lambda} = \frac{1}{2^{1/3}} \int_s^{d_0(z)} \frac{d\lambda}{\lambda^{5/3}(\lambda + 1)^{1/3}}.$$

Inverting this function we obtain the estimate

$$d_0(F_t(z)) \leq V(t) = \frac{d_0(z)}{\left[\dfrac{2^{4/3}}{3} td_0(z)^{2/3} + (d_0(z) + 1)^{2/3} \right]^{3/2} - d_0(z)}.$$

The latter inequality is equivalent to the estimate

$$|F_t(z)| \leq \frac{|z|}{\left[\dfrac{(2|z|)^{4/3}}{3} t + 1 \right]^{3/4}}.$$

Note that one can calculate F_t directly by solving the Cauchy problem and obtain

$$|F_t(z)|^2 = \frac{x^2}{\left(\frac{4}{3}x^{4/3}t + 1\right)^{3/2}} + \frac{y^2}{\left(\frac{4}{3}y^{4/3}t + 1\right)^{3/2}}.$$

Thus for $x = y$ we obtain

$$|F_t(z)| = \frac{|z|}{\left[\frac{(2|z|)^{4/3}}{3}t + 1\right]^{3/4}}.$$

So the rate of (nonexponential) convergence we have obtained is sharp.

Remark 4.7.2. We saw above (see Sections 4.4 and 4.5) that a similar phenomenon is impossible for holomorphic mappings:

Namely, *if a flow of holomorphic self-mappings converges locally uniformly to an* **interior stationary point** *then the convergence must be of exponential type.*

These examples and Proposition 4.7.1 above motivate the following definitions.

Definition 4.7.2 *Let $S = \{F_t\}_{t \geq 0}$ be a flow with a stationary (or sink) point $\tau \in \overline{\Delta}$. We will say that the asymptotic behavior of S at τ is of order not less than $\alpha > 0$ if there is a function $\omega \in \mathcal{M}(0, \infty)$ such that*

$$\liminf_{s \to 0^+} \left\{ \frac{\omega(s)}{s^{1/\alpha}} \right\} > 0 \tag{4.7.9}$$

and

$$d_\tau(F_t(z)) \leq \frac{1}{\left(1 + \frac{t}{\alpha}\omega(d_\tau(z))\right)^\alpha} d_\tau(z) \tag{4.7.10}$$

for all $z \in \Delta$ and $t \geq 0$.

Definition 4.7.3 *We will say that the asymptotic behavior of S at τ is of ω-exponential type if there is a decreasing function $\omega \in \mathcal{M}(0, \infty)$ such that*

$$d_\tau(F_t(z)) \leq \exp\left(-t\omega(d_\tau(z))\right) d_\tau(z) \tag{4.7.11}$$

for all $z \in \Delta$ and $t \geq 0$.

In particular, if ω can be chosen to be a positive constant a then we will say that S has a global uniform rate of exponential convergence:

$$d_\tau(F_t(z)) \leq \exp(-ta)d_\tau(z). \tag{4.7.12}$$

The following assertion is a consequence of Proposition 4.7.1.

Proposition 4.7.2 *Let $S = \{F_t\}_{t \geq 0}$ be a flow generated by $f \in \mathcal{GN}_\rho(\Delta)$ with a null (or sink) point $\tau \in \overline{\Delta}$. Then the asymptotic behavior of S at τ is of order not less than $\alpha > 0$ if and only if there exists an appropriate lower bound $\omega \in \mathcal{M}(0, \infty)$ for f such that*

$$(**) \qquad\qquad \frac{\omega(s)}{s^{1/\alpha}} \text{ is decreasing on } (0, \infty).$$

Proof. We first observe that condition (4.7.10) with some $\omega \in \mathcal{M}(0, \infty)$ satisfying (4.7.9) is equivalent to the same condition with a function $\omega_1 \in \mathcal{M}(0, \infty)$ which satisfies both (4.7.9) and (**). Indeed, for a given $\omega \in \mathcal{M}(0, \infty)$ define a function $\mu : (0, \infty) \mapsto (0, \infty)$ by

$$\mu(s) = \inf \left\{ \frac{\omega(l)}{l^{1/\alpha}} : l \in (0, s] \right\}, \quad s > 0.$$

It is clear that $\mu(s)$ is decreasing. Setting now $\omega_1(s) = s^{1/\alpha} \cdot \mu(s)$ we clearly see that ω_1 satisfies (4.7.9) and that $\omega_1(s) \le \omega(s)$. Hence

$$\int\limits_{0+} \frac{ds}{\omega_1(s)s}$$

is divergent and $\omega_1 \in \mathcal{M}(0, \infty)$. Then the inequality

$$\frac{1}{\left[1 + \frac{t}{\alpha}\omega(s)\right]^\alpha} \le \frac{1}{\left[1 + \frac{t}{\alpha}\omega_1(s)\right]^\alpha}$$

proves our claim.

Thus we can assume for the rest of the proof that ω satisfies (**). It remains to be shown that ω is an appropriate lower bound for f.

Indeed, defining $\Omega : (0, d_\tau(z)] \mapsto [0, \infty)$ by (1.7) and using (**) we have

$$
\begin{aligned}
\Omega(s) &= \int\limits_{s}^{d_\tau(x)} \frac{d\lambda}{\omega(\lambda)\lambda} = \int\limits_{s}^{d_\tau(x)} \frac{\lambda^{\frac{1}{\alpha}} \, d\lambda}{\omega(\lambda)\lambda^{\frac{1}{\alpha}+1}} \\
&\le \frac{[d_\tau(z)]^{\frac{1}{\alpha}}}{\omega(d_\tau(z))} \int\limits_{s}^{d_\tau(z)} \frac{d\lambda}{\lambda^{\frac{1}{\alpha}+1}} \\
&= \frac{\alpha}{\omega(d_\tau(z))} \left[s^{-\frac{1}{\alpha}} (d_\tau(z))^{\frac{1}{\alpha}} - 1 \right].
\end{aligned}
$$

Inverting this expression we obtain

$$V(t) := \Omega^{-1}(t) \le \frac{1}{\left(1 + \frac{t}{\alpha}\omega(d_\tau(z))\right)} d_\tau(z) := W(t).$$

It is clear that the function $W(t)$ satisfies all the conditions of Proposition 4.7.1. This completes the proof of Proposition 4.7.2. $\square$

Corollary 4.7.1 *Let $S = \{F_t\}_{t\ge0}$ be a flow generated by $f \in \mathcal{GN}_\rho(\Delta)$ with a null (or sink) point $\tau \in \overline{\Delta}$. Then:*

(i) The asymptotic behavior of S at τ is of ω-exponential type if and only if

$$\inf \left\{ \omega^\sharp(l) : l \in (0, s] \right\} > 0, \quad s > 0. \tag{4.7.13}$$

(ii) The flow S has a global uniform rate of exponential convergence if and only if

$$\omega^{\sharp}(s) \geq a \qquad (4.7.14)$$

for some a > 0.

Indeed, in both cases (i) and (ii) there is one function $\omega \in \mathcal{M}(0, \infty)$ such that the asymptotic behavior of S at τ is of order not less than α for all positive α. In case (i), ω can be chosen to be

$$\omega(s) := \inf\left\{\omega^{\sharp}(l) : l \in (0, s]\right\} > 0, \quad s > 0,$$

while in case (ii) ω can be chosen to be a constant a.

The following example shows that *for a semigroup of ρ-nonexpansive* (**but not holomorphic!**) *mappings an asymptotic behavior of ω-exponential type does not imply, in general, a global uniform rate of exponential convergence.*

Example 4. Define a continuous mapping $f : \Delta \mapsto \mathbb{C}$ by the following formula:

$$f(x + iy) = x(1 - x)^2 + iy(1 - y)^2.$$

Since $\mathrm{Re}\, f(z)\bar{z} \geq 0$ for all $z = x + iy \in \Delta$, it follows that f is a generator of a semigroup $S = \{F_t\}_{t \geq 0}$ of ρ-nonexpansive mappings such that each disk $\Delta_r = \{z \in \mathbb{C} : |z| < r < 1\}$ is F_t-invariant. Setting $\tau = 0$ and $z^* = \dfrac{z}{|z|^2(1 - |z|^2)}$, we have

$$\omega^{\sharp}(s) = \inf_{d_0(z)=s} \mathrm{Re}\, f(z)\bar{z}^* = \inf_{x^2+y^2=\frac{s}{s+1}} \frac{x^2(1-x)^2 + y^2(1-y)^2}{(x^2+y^2)(1-x^2-y^2)}.$$

It is easy to see that $\lim\limits_{s \to 0^+} \omega^{\sharp}(s) = 1$ while $\omega^{\sharp}(s) \to 0$ as $s \to \infty$ (take, for example, $y = 0$ and $x = \sqrt{\dfrac{s}{s+1}} \to 1$).

Remark 4.7.3 For holomorphic mappings, however, condition (4.7.14) holds for some $a > 0$ whenever condition (4.7.13) holds for a decreasing positive function ω. In other words, *for holomorphic flows any convergence of ω-exponential type implies global uniform exponential convergence.*

To explain this phenomenon in terms of lower bounds we observe that for a holomorphic generator with an interior null point the function

$$\omega_b(s) = \inf_{d_\tau(z) \leq s} 2\, \mathrm{Re}\, f(z)\overline{z^*} \leq \omega^{\sharp}(s) \qquad (4.7.15)$$

is bounded from below by a positive number, while for a boundary sink point the function ω_b is just a constant.

In both cases (interior stationary point or boundary sink point) the asymptotic behavior of a flow generated by $f \in \mathcal{G}\,\mathrm{Hol}(\Delta)$ is completely determined by the value $\omega^{\sharp}(0) := \liminf\limits_{s \to 0^{+}} \omega^{\sharp}(s)$ which is related to the value of derivative of f at its null point (for the interior case) or the angular derivative (for the boundary case).

Proposition 4.7.3 *Let $f \in \mathcal{G}\,\mathrm{Hol}(\Delta)$ and let $\{F_t\}_{t \geq 0}$ be the flow generated by f. If for some point $\tau \in \overline{\Delta}$ there is a decreasing function $\omega : (0, \infty) \mapsto (0, \infty)$ such that*

$$d_\tau(F_t(z)) \leq e^{-t\omega(d_\tau(z))}\, d_\tau(z), \quad z \in \Delta,\ t \geq 0, \tag{4.7.16}$$

then there exists a number $\mu > 0$ such that

$$d_\tau(F_t(z)) \leq e^{-t\mu} d_\tau(z), \quad z \in \Delta,\ t \geq 0. \tag{4.7.17}$$

Moreover,

(i) if $\tau \in \Delta$, then μ can be chosen as $\mu = \omega_\flat(0)/4$, but μ cannot be larger than $\omega_\flat(0)$ $(= \lim\limits_{s \to 0^{+}} \omega_\flat(s))$;

(ii) if $\tau \in \partial\Delta$, then the maximal μ for which (4.7.17) holds is exactly $\omega_\flat(0)$, that is $0 < \mu \leq \omega_\flat(0)$.

The proof of this Proposition is based on the following two lemmata.

Lemma 4.7.1 *Let $f \in \mathcal{G}\,\mathrm{Hol}(\Delta)$ with $f(0) = 0$, and let $\omega_\flat$ and $\omega^{\sharp}$ be defined by (4.1.4) and (4.1.5), respectively. Then:*

(i) $\omega^{\sharp}(0) = \omega_\flat(0) = 2\,\mathrm{Re}\,f'(0) := 2\nu$;

(ii) $\nu/2 \leq \omega_\flat(s) \leq 2\nu$.

Proof. First we show that

$$\omega^{\sharp}(0) \leq 2\nu, \tag{4.7.18}$$

where

$$\nu = \mathrm{Re}\,f'(0).$$

Since in our case $\tau = 0$, we have

$$\mathrm{Re}\,f(z)\overline{z^*} = \frac{1}{|z|^2(1 - |z|^2)}\,\mathrm{Re}\,f(z)\bar{z}.$$

Now fixing $\zeta \in \partial\Delta$, we set $z = r\zeta$, where $r \in (0,1)$. Then we obtain

$$\mathrm{Re}\,f(z)\overline{z^*} = \mathrm{Re}\,\frac{1}{1 - r^2}\frac{1}{r}\,f(r\zeta) \cdot \bar{\zeta}.$$

Therefore

$$\omega^{\sharp}(s) \leq 2\,\mathrm{Re}\,\frac{1}{1 - r^2}\frac{1}{r}\,f(r\zeta) \cdot \bar{\zeta}, \quad \text{where } r^2 = |z|^2 = \frac{s}{s+1}.$$

Letting s (hence, r) tend to zero we obtain

$$\omega^{\sharp}(0) \leq 2\,\mathrm{Re}\,f'(0).$$

On the other hand, it follows by Harnack inequality that for all $z \in \Delta$

$$\operatorname{Re} f(z)\bar{z} \geq \nu|z|^2 \frac{1-|z|}{1+|z|}.$$

This implies that

$$
\begin{aligned}
2 \operatorname{Re} f(z)\overline{z^*} &= \frac{2}{|z|^2(1-|z|^2)} \operatorname{Re} f(z)\bar{z} \\
&\geq \frac{2\nu|z|^2}{|z|^2(1-|z|^2)} \cdot \frac{1-|z|}{1+|z|} = \frac{2\nu}{(1+|z|)^2}.
\end{aligned}
$$

Hence

$$
\begin{aligned}
\omega_b(s) &\geq \inf_{d_\tau(z)\leq s} \frac{2\nu}{(1+|z|)^2} = \inf_{|z|^2 \leq \frac{s}{s+1}} \frac{2\nu}{(1+|z|)^2} \\
&= \frac{2\nu}{\left(1+\sqrt{\frac{s}{1+s}}\right)^2} \quad (\geq \nu/2).
\end{aligned} \tag{4.7.19}
$$

Letting s tend to 0^+ in (4.7.19) we see that $\omega_b(0) \geq 2\nu$. Comparing the latter inequality with (4.7.18) and (4.7.15) we obtain (i). On the other hand, substituting now in (4.7.19) $\nu = \omega_b(0)/2$, we obtain assertion (ii) and we are done. $\square$

Suppose now that $\mathcal{G}\operatorname{Hol}(\Delta)$ has a null point different from zero, say, $f(\tau) = 0$, $\tau \in \Delta$, $\tau \neq 0$.

Note that if the automorphism M_τ of Δ is given by

$$M_\tau(z) = \frac{\tau - z}{1 - z\bar{\tau}},$$

then the following equality holds:

$$d_\tau\left(M_\tau(z)\right) = (1 - |\tau|^2)d_0(z). \tag{4.7.20}$$

Now let us consider the flow $\{G_t\}_{t\geq 0} \subset \operatorname{Hol}(\Delta)$ defined by

$$G_t = M_\tau \circ F_t \circ M_\tau \tag{4.7.21}$$

and let $g \in \mathcal{G}\operatorname{Hol}(\Delta)$ be its generator, i.e.,

$$g(z) = -\frac{\partial}{\partial t}G_t(z)|_{t=0^+} = [(M_\tau)'(z)]^{-1} f(M_\tau(z)). \tag{4.7.22}$$

Then $G_t(0) = 0$ for all $t \geq 0$ and $g(0) = 0$.

Lemma 4.7.2 *The functions* $\omega^\#(s)$ *and* $\omega_b(s)$ *are invariant under the transformations (4.7.21) and (4.7.22).*

Proof. We have seen already in (4.7.7) and (4.7.8) that

$$\frac{\partial}{\partial t}\left[d_\tau(F_t(z))\right]_{t=0^+} = -2d_\tau(z)\operatorname{Re} f(z)\overline{z^*}$$

$$= -2\frac{d_\tau(z)}{1-\sigma(z,\tau)}\operatorname{Re}\left[f(z)\left(\frac{\bar{z}}{1-|z|^2}-\frac{\bar{\tau}}{1-z\bar{\tau}}\right)\right]. \quad (4.7.23)$$

On the other hand, by (4.7.20) and (4.7.21) we have

$$\frac{\partial}{\partial t}\left[d_\tau(F_t(z))\right]_{t=0^+} = \frac{\partial}{\partial t}\left[d_\tau(M_\tau G_t(w))\right]_{t=0^+}$$

$$= -\frac{\partial}{\partial t}\left[(1-|\tau|^2)d_0(G_t(w))\right]_{t=0^+}. \quad (4.7.24)$$

Since

$$\frac{\partial}{\partial t}\left[d_0(G_t(w))\right]_{t=0^+} = -\frac{d_0(w)}{|w|^2}2\operatorname{Re}\left(g(w)\frac{\bar{w}}{1-|w|^2}\right)$$

$$= -\frac{d_\tau(z)}{(1-|\tau|^2)|w|^2}2\operatorname{Re}\left(g(w)\frac{\bar{w}}{1-|w|^2}\right),$$

we obtain from (4.7.24) and (4.7.23) the required equality

$$\frac{1}{1-\sigma(z,\tau)}\operatorname{Re}\left[f(z)\left(\frac{\bar{z}}{1-|z|^2}-\frac{\bar{\tau}}{1-z\bar{\tau}}\right)\right] = \frac{1}{|w|^2}\operatorname{Re}\left[g(w)\frac{\bar{w}}{1-|w|^2}\right].$$

The Lemma is proved. □

Let the flow $\{G_t(z)\}_{t\geq 0}\subset\operatorname{Hol}(\Delta)$ and its generator $g\in\mathcal{G}\operatorname{Hol}(\Delta)$ be defined by (4.7.21) and (4.7.22).

By Lemmata 4.7.2 and 4.7.1 we have

$$\inf_{d_0(w)=s}\operatorname{Re} g(w)\overline{w^*}\geq\frac{\omega_b(0)}{4}=\frac{\nu}{2}.$$

Then by Corollary 4.7.1(ii) we have

$$d_0(G_t(w))\leq d_0(w)\exp(\nu/2)\quad\text{for all }w\in\Delta.$$

Finally, setting $w=M_\tau(z)$ and using (4.7.20) we conclude that

$$d_\tau(F_t(z))\leq d_\tau(z)\exp(\nu/2). \quad (4.7.25)$$

This enable us to point out the following rates of convergence.

Proposition 4.7.4 Let $\{F_t\}_{t>0}$ be a flow generated by $f\in\mathcal{G}\operatorname{Hol}(\Delta)$ and let $f(\tau)=0$, $\tau\in\Delta$. Then the following estimates are equivalent:

(i) $d_\tau(F_t(z)) \leq d_\tau(z) \exp(-t\mu)$, $z \in \Delta$, $t \geq 0$;

(ii) $|M_\tau(F_t(z))| \leq |M_\tau(z)| \cdot \exp\left(-\mu\dfrac{1 - |M_\tau(z)|^2}{2}t\right)$, $z \in \Delta$, $t \geq 0$;

(iii) $|M_\tau(F_t(z))| \leq |M_\tau(z)| \cdot \exp\left(-\nu\dfrac{1 - |M_\tau(z)|}{1 + |M_\tau(z)|}t\right)$, $z \in \Delta$, $t \geq 0$,

where the numbers μ in (i) and (ii) can be chosen to be one and the same such that $0 \leq \nu/2 \leq \mu \leq 2\nu$ and ν in (iii) is defined by

$$\nu = \frac{1}{2}\omega_b(0) = \frac{1}{2}\omega^\sharp(0) = \operatorname{Re} f'(\tau). \tag{4.7.26}$$

Proof. First we note that inequalities (ii) and (iii) are equivalent to the following ones:

(ii^*) $|G_t(w)| \leq |w| \cdot \exp\left(-\mu\dfrac{1 - |w|^2}{2}t\right)$, $t \geq 0$,

(iii^*) $|G_t(w)| \leq |w| \cdot \exp\left(-\nu\dfrac{1 - |w|}{1 + |w|}t\right)$, $t \geq 0$,

where $w = M_\tau(z) \in \Delta$ and the flow $\{G_t\}_{t\geq 0}$ is defined by (4.7.21). First let us suppose that estimate (i) holds. By using (4.7.20) for this flow we have

$$d_0(G_t(w)) \leq d_0(w) \exp(-t\mu).$$

Rewriting the latter inequality in the form

$$\frac{|G_t(w)|^2}{1 - |G_t(w)|^2} \leq \frac{|w|^2}{1 - |w|^2} \cdot \exp(-t\mu),$$

we obtain by direct calculations

$$|G_t(w)|^2 \leq |w|^2 \frac{1}{|w|^2 + (1 - |w|^2)\exp(-t\mu)} \leq |w|^2 \cdot \exp\left(-t\mu(1 - |w|^2)\right).$$

which coincides with (ii^*).

Now we will assume that inequality (ii) (and hence (ii^*)) holds. Differentiating both sides of this inequality with respect to t at $t = 0^+$ we obtain

$$-\frac{1}{|w|}\operatorname{Re} g(w)\bar{w} \leq -|w|\mu\frac{1 - |w|^2}{2}. \tag{4.7.27}$$

This implies that $\omega^\sharp(s) \geq \mu$. Thus the function $\omega(s) \equiv \mu$ is an appropriate lower bound and the implication $(ii) \Rightarrow (i)$ follows by Proposition 4.7.2.

Let us suppose now again that inequality (ii) (hence, (ii^*) and (4.7.27)) holds with some number $\mu > 0$. Setting in (4.7.27) $w = r\zeta$, $\zeta \in \partial\Delta$, $r \in (0, 1)$ and letting r to zero (cf., the proof of Lemma 4.7.1) we obtain $\operatorname{Re} g'(0) \geq \mu/2 > 0$. A direct calculation shows that

$$g'(0) = [(M_\tau)'(0)]^{-1} f'(\tau) (M_\tau)'(0) = f'(\tau).$$

and so $\nu > 0$. Therefore, again by Harnack inequality, we have

$$\operatorname{Re} g(w)\bar{w} \geq \operatorname{Re} g'(0)\frac{1 - |w|}{1 + |w|} \geq \nu|w|^2\frac{1 - |w|}{1 + |w|}.$$

On the other hand,

$$\frac{\partial \ln |G_t(w)|}{\partial t} = -\frac{1}{|G_t(w)|^2} \operatorname{Re}\left[g(G_t(w))\overline{G_t(w)}\right].$$

Also it follows by the Schwarz Lemma that $|G_t(w)| \leq |w|$. Thus we have

$$\frac{\partial \ln |G_t(w)|}{\partial t} \leq -\nu\frac{1 - |w|}{1 + |w|}.$$

Integrating this inequality we obtain the following estimate

$$\ln |G_t(w)| - \ln |w| \leq -\nu\frac{1 - |w|}{1 + |w|}t,$$

which implies (iii^*).

Finally, if condition (iii^*) holds then differentiating it with respect to t at $t = 0^+$ we obtain

$$\operatorname{Re} g(w)\bar{w} \geq \frac{\nu}{(1 + |w|)^2} \geq \frac{\nu}{4} > 0.$$

Thus $\omega^\#(0) > 0$ and the result follows. $\square$

Let us turn now to the case when $f \in \mathcal{G} \operatorname{Hol}(\Delta)$ has no null points in Δ.

As a matter of fact, if $S = \{F_t\}_{t \geq 0}$ converges to a boundary sink point $\tau \in \partial\Delta$ with a rate of convergence of exponential type,

$$d_\tau(F_t(z)) \leq \exp\left(-t\omega(d_\tau(z))\right) \cdot d_\tau(z),$$

where $\omega \in \mathcal{M}(0, \infty)$ is a decreasing function, then this estimate can be improved as follows:

$$d_\tau(F_t(z)) \leq \exp\left(-t\omega(0)\right)d_\tau(z),$$

where $\omega(0) := \lim_{s \to 0^+} \omega(s)$.

In other words, we claim that if the inequality

$$\omega^\#(s) \geq \omega(s)$$

holds for a decreasing ω then the stronger inequality

$$\omega^\#(s) \geq \omega(0)$$

also holds. In particular, this property holds for the function $\omega = \omega_b$. This implies, in turn, that ω_b is actually constant: $\omega_b(s) = \omega_b(0) = \beta$ for all $s \in (0, \infty)$ and is equal to the angular derivative of f at the point $\tau \in \partial\Delta$. Moreover, this number β gives the best rate of exponential convergence of $S = \{F_t\}_{t \geq 0}$.

Proposition 4.7.5 *Let the function* $f \in \text{Hol}(\Delta, \mathbb{C})$ *be a generator of a semigroup* $S = \{F_t\}_{t \geq 0}$ *of holomorphic self-mappings of* Δ. *Suppose that* f *has no null-point in* Δ *and that* $\tau \in \partial\Delta$ *is the boundary sink point for* S. *Then the following are equivalent:*

(i) *the asymptotic behavior of* S *at* τ *is of* ω-*exponential type;*
(ii) *there is a positive number* γ *such that*

$$d_\tau(F_t(z)) \leq d_\tau(z) \exp(-t\gamma), \quad z \in \Delta \text{ and } t \geq 0.$$

Moreover, the maximal γ *which satisfies condition (ii) is*

$$\beta = \angle f'(\tau) = 2 \inf_{z \in \Delta} \text{Re } f(z)\overline{z^*}.$$

Proof. Let condition (i) hold, i.e., for some decreasing function $\omega \in \mathcal{M}(0, \infty)$,

$$d_\tau(F_t(z)) \leq d_\tau(z) \exp(-t\omega(d_\tau(z)))$$

or explicitly,

$$\frac{|1 - F_t(z)\bar{\tau}|^2}{1 - |F_t(z)|^2} \leq \frac{|1 - z\bar{\tau}|^2}{1 - |z|^2} \exp(-2t\omega(d_\tau(z))).$$

This is equivalent to the inequality

$$\frac{|1 - F_t(z)\bar{\tau}|^2}{|1 - z\bar{\tau}|^2} \leq \frac{1 - |F_t(z)|^2}{1 - |z|^2} \exp(-2t\omega(d_\tau(z))). \tag{4.7.28}$$

Once again it follows by the Julia–Carathéodory Theorem that for fixed $t \geq 0$,

$$\delta(F_t) := \liminf_{z \to \tau} \frac{1 - |F_t(z)|}{1 - |z|}$$

$$= \angle[F_t]'(\tau) := \lim_{z \to \tau} \frac{1 - F_t(z)\bar{\tau}}{1 - z\bar{\tau}}, \tag{4.7.29}$$

where that last limit is taken along an nontangential approach region at τ.

Let us denote

$$\omega(0) = \lim_{s \to 1^-} \omega(d_\tau(s\tau)).$$

Thus setting $z = s\tau$ in (4.7.28) and letting s tend to 1^- we obtain

$$\delta^2(F_t)) \leq \delta(F_t) \exp(-2t\omega(0)),$$

or

$$\delta(F_t) \leq \exp(-t\gamma),$$

where we set $\gamma = 2\omega(0)$.

Now by using Julia's Lemma we obtain the implication $(i) \Rightarrow (ii)$. The converse implication can be shown by differentiating the inequality in (ii) at $t = 0^+$. Namely, we obtain

$$\mathrm{Re}\, f(z)\overline{z^*} \geq \frac{\gamma}{2} > 0.$$

So one can set $\omega(s) \equiv \gamma/2$ and the asymptotic behavior of S at τ is seen to be of exponential type.

In addition, it follows by Proposition 4.6.2 that $\gamma \leq \beta := \angle f'(\tau)$.

Finally, we observe that

$$\lim_{s \to 1^-} \mathrm{Re}\, f(s\tau)\overline{(s\tau)^*} \;=\; \lim_{r \to 1^-} \mathrm{Re}\, f(s\tau) \left(\frac{s\overline{\tau}}{1 - s^2} - \frac{\overline{\tau}}{1 - s} \right)$$

$$=\; \lim_{s \to 1^-} \frac{\mathrm{Re}\, f(s\tau)\overline{\tau}}{s - 1} \frac{1}{1 + s} = \beta/2,$$

and the assertion is proved. $\square$

Chapter 5

Dynamical approach to starlike and spirallike functions

This chapter is devoted to showing some relationships between semigroups and the geometry of domains in the complex plane. Mostly we will study those univalent (one-to-one correspondence) functions on the unit disk whose images are starshaped or spiralshaped domains. Several important aspects, however, had to be omitted, e.g. convex and close-to-convex functions (see, for example, [57, 55]), and other different classes of univalent functions. We have selected the forthcoming material according to the guiding principle that the demonstrated methods may be generalized to higher dimensions. For example, the celebrated Koebe One Quarter Theorem states that the image of a univalent function h on Δ normalized by the condition $h(0) = 0$ and $h'(0) = 1$ contains a disk of radius $\frac{1}{4}$. This theorem is no longer true at higher dimensions. Nevertheless, the dynamical approach analogues of the Koebe theorem have been recently established and used for subclasses of starlike (or spirallike) functions (see, for example [141, 109, 26, 56, 14]).

Our second objective is to study the dynamics of starshaped (or spiralshaped) domains when the origin is pushed out to the boundary. For example, a domain which is starshaped with respect to a point may fail to be starshaped with respect to another point. We will study *inter alia* some unified conditions which describe starlike (or spirallike) functions which are independent of the location of their null points.

5.1   Generators on biholomorphically equivalent domains

Although the studies in the previous chapters were carried out on the unit disk, one can translate them to any simply connected domain (which differs from $\mathbb{C}$) of the plane as the consequence of the Riemann Mapping Theorem. First we recall some notions and definitions in classical function theory.

Definition 5.1.1 *Let D be a domain in $\mathbb{C}$. A function $h \in \text{Hol}(D, \mathbb{C})$ is said to be univalent on D if for each pair of distinct points z_1 and z_2 in D we have $h(z_1) \neq h(z_2)$.*

The set of all univalent functions in a domain $D \subset \mathbb{C}$ will be denoted by $\text{Univ}(D)$.

For $h \in \text{Univ}(D)$ one can define the inverse mapping $h^{-1} : \Omega \mapsto D$, where $\Omega = h(D)$. The content of the Open Mapping Theorem (see, for example, [122]) is that $\Omega = h(D)$ is also a domain (open connected subset) in $\mathbb{C}$. In addition, $h^{-1} \in \text{Hol}(\Omega, D)$. In other words, h is one-to-one and h^{-1} is also holomorphic on $h(D)$. In this case f is also called a (globally) biholomorphic mapping on D.

A mapping $h \in \text{Hol}(D, \mathbb{C})$ is said to be locally biholomorphic on D if for each point $z \in D$ there is a neighborhood $V \subset D$, of this point such that $h \in \text{Univ}(V)$. It is well known that $h \in \text{Hol}(D, \mathbb{C})$ is locally biholomorphic on D if and only if $h'(z) \neq 0$ everywhere (see, for example, [122, 128]).

Two domains D and Ω in $\mathbb{C}$ are called biholomorphically (or conformally) equivalent if there exists $h \in \text{Univ}(D)$ such that $\Omega = f(D)$.

The fundamental **Riemann Mapping Theorem** states that *every simply connected domain Ω in $\mathbb{C}$ (but not $\mathbb{C}$ itself) is biholomorphically equivalent to the open unit disk Δ in $\mathbb{C}$. Moreover, for each $a \in \Omega$ there is a unique $h \in \text{Univ}(\Delta)$ with $\Omega = f(\Delta)$ such that $h(0) = a$ and $h'(0) > 0$.*

For the special case when $D = \Delta$ is the open unit disk in $\mathbb{C}$, the subset of $\text{Univ}(D)$ normalized by the conditions

$$h(0) = 0 \text{ and } h'(0) = 1$$

will be denoted by $S(\Delta)$. This notation conforms to the one used in the classical geometric function theory. In this case we simply write

$$S\left(= S(\Delta)\right) = \{h \in \text{Univ}(\Delta) : h(0) = 0 \text{ and } h'(0) = 1\}.$$

In other words, the class $S \subset \text{Hol}(\Delta, \mathbb{C})$ consists of all the mappings $h \in \text{Univ}(\Delta)$ such that h has the following Taylor series at the origin:

$$h(z) = z + \sum_{k=2}^{\infty} a_k z^k.$$

Thus, referring to some geometrical properties of a simply connected domain Ω containing the origin, if we are permitted to shrink or expand it we can find a function $h \in S (= S(\Delta))$ for which the domain $\widetilde{\Omega} = h(\Delta)$ is similar to Ω. (Of course, we can translate a domain, if necessary, so that the origin would be its interior point). However, in this way we may sometimes loose some features of the dynamical transformation of a domain Ω if its geometrical characteristics are related to a certain given fixed point in $\mathbb{C}$. In particular, it happens if such a point lies on the boundary of Ω.

The following simple assertion is the key to our further considerations.

Proposition 5.1.1 (Main Lemma) *Let D and Ω be two domains in $\mathbb{C}$, such that $\Omega = h(D)$ for some biholomorphic (univalent mapping) h. Then there is a linear invertible operator T on the space $\mathrm{Hol}(\Omega, \mathbb{C})$ onto the space $\mathrm{Hol}(D, \mathbb{C})$, which takes the set $\mathcal{G} \mathrm{Hol}(\Omega) \subset \mathrm{Hol}(\Omega, \mathbb{C})$ onto the set $\mathcal{G} \mathrm{Hol}(D)$ (i.e., $\mathcal{G} \mathrm{Hol}(D) = T(\mathcal{G} \mathrm{Hol}(\Omega))$). Moreover, such an operator $T : \mathcal{G} \mathrm{Hol}(\Omega) \longleftrightarrow G Hol(D)$ can be given by the formula:*

$$T(\varphi)(\cdot) = [h'(\cdot)]^{-1} \varphi(h(\cdot)), \tag{5.1.1}$$

where $\varphi \in \mathcal{G} \mathrm{Hol}(\Omega)$.

Proof. Let $\varphi \in \mathcal{G} \mathrm{Hol}(\Omega)$ and let $S_\varphi = \{\phi_t\}_{t \geq 0}$ be the semigroup of holomorphic self-mappings on Ω generated by φ. Then it is clear that the family $\{F_t\}_{t \geq 0}$ defined by:

$$F_t = h^{-1} \circ \phi_t \circ h \tag{5.1.2}$$

is a semigroup of holomorphic self-mappings of D, which is continuous with respect to $t \geq 0$. By $f \in \mathcal{G} \mathrm{Hol}(D, \mathbb{C})$ we denote the generator of this semigroup, i.e.,

$$f(z) = \lim_{t \to 0^+} \frac{z - F_t(z)}{t} = -\frac{dF_t(z)}{dt} \Big|_{t=0}. \tag{5.1.3}$$

Then substituting formula (5.1.2) into (5.1.3) and using the chain rule we obtain:

$$
\begin{aligned}
f(z) &= -\left[(h^{-1})'(\phi_t(h(z))) \cdot \frac{d\phi_t(h(z))}{dt} \right]_{t=0} = (h^{-1})'(h(z)) \cdot \varphi(h(z)) \\
&= [h'(z)]^{-1} \cdot \varphi(h(z)), \quad z \in D,
\end{aligned}
$$

i.e., $f = T\varphi \in \mathcal{G} \mathrm{Hol}(D)$, where T is defined by (5.1.1). Since $h'(z) \neq 0$ it is clear that T is a well defined linear action on $\mathrm{Hol}(\Omega, \mathbb{C})$ into $\mathrm{Hol}(D, \mathbb{C})$. In addition, since h is biholomorphic on D the operator T is invertible and for each $f \in \mathrm{Hol}(D, \mathbb{C})$, its inverse $T^{-1} : \mathrm{Hol}(D, \mathbb{C}) \mapsto \mathrm{Hol}(\Omega, \mathbb{C})$ can be defined by

$$T^{-1}(f)(w) = [(h^{-1})'(w)]^{-1} \cdot f(h^{-1}(w)). \tag{5.1.4}$$

Repeating the above considerations we can see, that if f in (5.1.4) belongs to $\mathcal{G} \mathrm{Hol}(D)$, then $\varphi = T^{-1}(f)$ belongs to $\mathcal{G} \mathrm{Hol}(\Omega)$, that is T maps $\mathcal{G} \mathrm{Hol}(\Omega)$ onto $\mathcal{G} \mathrm{Hol}(D)$. $\square$

Remark 5.1.1. It follows by formula (5.1.2) that if $\varphi \in \text{aut}(\Omega)$, then $f = T\varphi \in \text{aut}(D)$ and conversely. Thus T maps the set $\text{aut}(\Omega)$ of all complete vector fields on Ω onto the set $\text{aut}(D)$ of all complete vector fields on D.

If, in particular, $D = \Omega$ then $\text{aut}(D)$ is an invariant subset of $\text{Hol}(D, \mathbb{C})$ for operator T defined by (5.1.1).

Remark 5.1.2. If $\varphi \in \mathcal{G}\,\text{Hol}(\Omega)$ has a null point $b \in \Omega$, then so does $f = T\varphi$, and $a = h^{-1}(b)$ is a null point of f. Indeed,

$$f(a) = [h'(a)]^{-1} \varphi(h(a)) = [h'(a)]^{-1} \cdot \varphi(b) = 0.$$

In addition, note that by direct calculations we obtain:

$$f'(a) = \varphi'(b). \tag{5.1.5}$$

Therefore a is an attractive fixed point of the semigroup $\{F_t\} = S_f$ if and only if b is an attractive fixed point of the semigroup $\{\phi_t\} = S_\varphi$.

Remark 5.1.3. By using Proposition 5.1.1 and Remark 5.1.2 we are able to easily obtain a description of the class $\mathcal{G}\,\text{Hol}(\Delta, \tau)$ of all semi-complete vector fields vanished in Δ, at a point $\tau \in \Delta$, which is a particular case of the representation that according to E. Berkson and H. Porta (see Section 4.6).

Indeed, let us set $D = \Omega = \Delta$ and $h = M_\tau \in \text{aut}(\Delta)$ for some $\tau \in \Delta$, where

$$M_\tau(z) = \frac{\tau - z}{1 - z\bar{\tau}}. \tag{5.1.6}$$

If $f \in \mathcal{G}\,\text{Hol}(\Delta, \tau)$ is a semi-complete vector field, such that $f(\tau) = 0$ for some $\tau \in \Delta$, then $\varphi = T^{-1}f$ is an element of $\mathcal{G}\,\text{Hol}(\Delta, 0)$ such that $\varphi(0) = 0$ (note that $h^{-1}(\tau) = 0$). Then φ has a representation: $\varphi(z) = z \cdot p_1(z)$ when $\text{Re}\,p_1(z) \geq 0$ for all $z \geq 0$. Then by Proposition 5.1.1 we obtain:

$$\begin{aligned}
f(z) = [h'(z)]^{-1} \cdot \varphi(h(z)) &= [h'(z)]^{-1} \cdot h(z) \cdot p(M_\tau(z)) \\
&= \frac{(1 - z\bar{\tau})^2}{|\tau|^2 - 1} \cdot \frac{\tau - z}{1 - z\bar{\tau}} \cdot p(M_\tau(z)).
\end{aligned}$$

Now, if we denote $q(z) = \dfrac{1}{1 - |\tau|^2} p(M_\tau(z))$ we obtain the desired representation:

$$f(z) = (z - \tau)(1 - z\bar{\tau})q(z), \tag{5.1.7}$$

where $\text{Re}\,q(z) \geq 0$ everywhere.

As a matter of fact, we know that representation (5.1.7) holds also for a null point free $f \in \mathcal{G}\,\text{Hol}(\Delta)$. In this case just $\tau \in \partial\Delta$.

Therefore we can try to use this representation to characterize different biholomorphic mappings on Δ and the geometric structure of their images.

5.2   Starlike and spirallike functions

Definition 5.2.1 *A set Ω in $\mathbb{C}$ is called starshaped (with respect to the origin) if given any $w \in \Omega$, the point tw belongs to Ω for every t with $0 < t \leq 1$. That is, if Ω contains w then it also contains the entire line segment joining w to the origin.*

If D is a domain in $\mathbb{C}$ then a function $h \in \text{Univ}(D)$ is said to be a starlike function on D if the image $\Omega = h(D)$ is a starshaped set (with respect to the origin).

In this definition the origin is in $\overline{\Omega}$. If, in particular, the origin belongs to Ω then we will say that h is **a starlike function with respect to an interior point**. In this case the function h has a null point τ in D. The image of the starlike function

$$f(z) = \frac{z}{(1 - 0.9z)^{0.7}(1 + 0.9iz)^{0.5}(1 - (0.7 + 0.4i)z)^{0.6}(1 + (0.56 + 0.7i)z)^{0.3}}$$

on the unit disk Δ is illustrated in Figure 5.1.

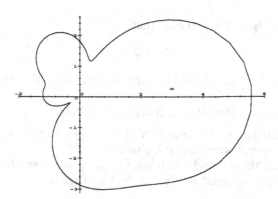

Figure 5.1: A starshaped domain $(0 \in \Omega)$.

If the origin belongs to the boundary $\partial\Omega$ of Ω we say that h is **starlike with respect to a boundary point (or fanlike on D)**. In this case there is a boundary point $\tau \in \partial D$ and a sequence $\{z_n\}_{n=1}^{\infty} \in D$, converging to τ such that

$$\lim_{n \to \infty} h(z_n) = 0.$$

The domain Ω in Figure 5.2 is the image of a fanlike (starlike with respect to a boundary point) on the unit disk Δ function

$$f(z) = \frac{(1 + z)^2}{(1 - 0.9z)^{0.5}(1 + 0.9iz)^{0.7}(1 - (0.7 + 0.5i)z)^{0.6}(1 + (0.6 + 0.67i)z)^{0.3}}.$$

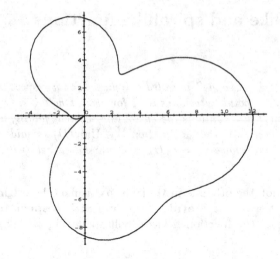

Figure 5.2: A starshaped domain $(0 \in \partial\Omega)$.

The set of all univalent functions on D which are starlike on D will be denoted by $\mathrm{Star}(D)$.

If, in addition, there is $\tau \in D$ such that

$$h(\tau) = 0,$$

then we will write $h \in \mathrm{Star}(D, \tau)$. Of course, in this case such a point τ is unique because

$$\mathrm{Star}(D, \tau) \subset \mathrm{Star}(D) \subset \mathrm{Univ}(D).$$

If $h \in \mathrm{Star}(D)$ has no null point in D (i.e., h is starlike with respect to a boundary point) we will write that $h \in \mathrm{Fan}(D)$.

Finally, keeping the classical notations, for the special case when $D = \Delta$ is the unit disk we will just write S^* for $\mathrm{Star}(\Delta, 0) \cap S$. That is

$$S^* = \{h \in \mathrm{Star}(\Delta) : h(0) = 0 \text{ and } h'(0) = 1\}.$$

The concept of univalent starlike functions was first introduced by Alexander [10] in 1915. In 1921 Nevanlinna [102] conducted a more detailed study of this class. In particular, the following characterization of the class $\mathrm{Star}(\Delta, 0)$ is due to him.

Proposition 5.2.1 *Let h be a univalent holomorphic mapping on the unit disk Δ such that*

$$h(0) = 0. \tag{5.2.1}$$

Then h is a starlike function on Δ if and only if

$$\mathrm{Re}\left[\frac{zh'(z)}{h(z)}\right] > 0, \quad z \in \Delta. \tag{5.2.2}$$

As we will see in the sequel, if $h \in \text{Hol}(\Delta, \mathbb{C})$ is locally biholomorphic, i.e., $h'(z) \neq 0$ everywhere, and satisfies (5.2.2) then it is necessarily univalent.

Furthermore, condition (5.2.2) leads to the study of other interesting subclasses of S. In particular, in 1936 Robertson [119] had introduced the class $S^*(\lambda)$ of starlike functions of order λ:

$$S^*(\lambda) = \left\{ h \in S^* : \text{Re} \left[\frac{zh'(z)}{h(z)} \right] \geq \lambda > 0, \quad z \in \Delta \right\}. \qquad (5.2.3)$$

In 1978 Wald [150] characterized starlike functions with respect to another center. Using our notions, his result can be reformulated in the following way.

Proposition 5.2.2 *Suppose that $h \in \text{Hol}(\Delta, \mathbb{C})$ is either of the form $h(z) = z + \sum_{k=2}^{\infty} a_k z^k$ or of the form $h(z) = \sum_{n=1}^{\infty} b_k z^k$ with $h'(\tau) = 1$ for some $\tau \in \Delta$. Then the function $h(z) - h(\tau)$ belongs to $\text{Star}(\Delta, \tau)$ if and only if $\text{Re}\, q(z) > 0$, where*

$$q(z) = \frac{h(z)}{h'(z)\,(z - \tau)(1 - z\bar{\tau})}, \quad z \in \Delta \setminus \{\tau\},$$

and

$$q(\tau) = 1 - |\tau|^2.$$

We will see below that Proposition 5.2.2 (as well as Proposition 5.2.1) can be easily obtained by using a different approach in more general settings. Different applications of this assertion are presented in Wald's thesis [150] (see also [57]).

Definition 5.2.2 *A set Ω in $\mathbb{C}$ is said to be spiralshaped (with respect to the origin) if there is a number $\mu \in \mathbb{C}$ with $\text{Re}\, \mu > 0$ such that for each $w \in \Omega$ and $t \geq 0$ the point $e^{-t\mu}w$ also belongs to Ω. (see Figure 5.3)*

If D is a domain in $\mathbb{C}$, then a univalent function $h \in \text{Hol}(D, \mathbb{C})$ is said to be spirallike on D if the closure of its image $\Omega = h(D)$ is a spiralshaped set.

Once again, if the origin belongs to Ω then we will say that h is a **spirallike function on D with respect to an interior point**. Otherwise (if the origin belongs to the boundary $\partial\Omega$ of Ω) we say that h is **spirallike with respect to a boundary point**.

The set of all biholomorphic mappings on D which are spirallike on D will be denoted by $\text{Spiral}(D)$. If, in addition, there is $\tau \in D$ such that

$$h(\tau) = 0,$$

then we will write $h \in \text{Spiral}(D, \tau)$. Note also that if in Definition 5.2.2 the number μ is a real positive number, then Ω is actually starshaped, i.e., $\text{Star}(D) \subset \text{Spiral}(D)$. Consequently, $\text{Star}(D, \tau) \subset \text{Spiral}(D, \tau)$.

If $h \in \text{Spiral}(D)$ has no null point in D (i.e., h is spirallike with respect to a boundary point) we will write that $h \in \text{Snail}(D)$. (We will see below that the class $\text{Snail}(D)$ is quite narrow: in particular, each $h \in \text{Univ}(\Delta)$ which has a biholomorphic extension to $\overline{\Delta}$ is, in fact, in $\text{Fan}(\Delta)$, i.e., $h(\Delta)$ is starshaped).

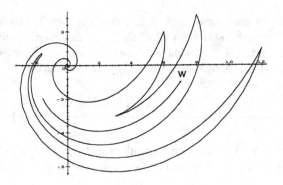

Figure 5.3: The spiralshaped domain $(0 \in \partial\Omega)$: $e^{-t\mu}w \in \Omega$ for each $w \in \Omega$.

It seems that the first occurrence of the class Spiral$(\Delta, 0) = \{h \in \text{Spiral}(\Delta) : h(0) = 0\}$ appeared when condition (5.2.2) was modified analytically by inserting the factor $e^{i\theta}$:

$$\text{Re}\left[e^{i\theta}\frac{zh'(z)}{h(z)}\right] > 0 \qquad (5.2.4)$$

(see Montel [101] and Špaček [137]). Actually, this definition is compatible with our Definition 5.2.2 (see also, Remark 5.2.2 below)

Proposition 5.2.3 *Let $h \in \text{Hol}(\Delta, \mathbb{C})$ have the form*

$$h(z) = z + \sum_{k=2}^{\infty} a_k z^k \qquad (5.2.5)$$

(i.e., $h(0) = 0$ and $h'(0) = 1$). Then $h \in \text{Spiral}(\Delta, 0)$ if and only if condition (5.2.4) holds for some $\theta \in (\pi/2, \pi/2)$.

Of course, Proposition 5.2.1 follows from Proposition 5.2.3 if we set there $\theta = 0$.

To prove the latter proposition as well as Proposition 5.2.2 we will need another more general assertion which is a direct consequence of the Main Lemma (Proposition 5.1.1).

Proposition 5.2.4 *Let $h \in \text{Hol}(D, \mathbb{C})$ be a spirallike (respectively, starlike) function on D. Then there exists a generator $f \in \mathcal{G}\,\text{Hol}(D)$ of a continuous flow on D such that h satisfies the following differential equation:*

$$\mu h(z) = h'(z)f(z), \quad z \in D, \qquad (5.2.6)$$

for some $\mu \in \mathbb{C}$ (respectively, $\mu \in \mathbb{R}$) with $\text{Re}\,\mu > 0$.

Proof. Let $\Omega = h(D)$ be a spiralshaped (respectively, starshaped) domain. Then for some $\mu \in \mathbb{C}$ (respectively, $\mu \in \mathbb{R}$) with $\text{Re}\,\mu > 0$ the mapping

$g = \mu I$, $(g(z) = \mu z)$, belongs to $\mathcal{G}\,\mathrm{Hol}(\Omega)$. Indeed, in this case the integral curve defined by the Cauchy problem

$$
\begin{cases}
\dfrac{\partial v(t,w)}{\partial t} + g(v(t,w)) = \dfrac{\partial v(t,w)}{\partial t} + \mu v(t,w) = 0, \\[2mm]
v(0,w) = w, \quad w \in \Omega,
\end{cases}
\tag{5.2.7}
$$

has the form $\{v(t,w) = e^{-\mu t}w,\ t \geq 0,\ w \in \Omega\}$ and belongs to Ω by Definition 5.2.2 (respectively, Definition 5.2.1). Then Proposition 5.1.1 implies that $f \in \mathrm{Hol}(D,\mathbb{C})$ defined by:

$$
f(z) = (Tg)(z) = \mu\,[h'(z)]^{-1}\,h(z)
\tag{5.2.8}
$$

is a generator of a flow on D, and we are done. $\square$

To establish a converse assertion we may omit the requirement on h to be univalent.

Proposition 5.2.5 *Suppose that $h \in \mathrm{Hol}(D)$ satisfies equation (5.2.6), with $\mu \in \mathbb{C}$, (respectively, $\mu \in \mathbb{R}$) $\mathrm{Re}\,\mu > 0$, and $f \in \mathcal{G}\,\mathrm{Hol}(D)$. Then the set $\Omega = h(D)$ is spiralshaped (respectively, starshaped). Moreover, if D is bounded and h has a null point $\tau \in D$ with $h'(\tau) \neq 0$, then h is univalent, hence spirallike (respectively, starlike) with respect to an interior point.*

Proof. Fix $w \in \Omega$ and find $z \in D$ such that $h(z) = w$. Since $f \in \mathcal{G}\,\mathrm{Hol}(D)$ generates a continuous flow $S_f = \{F_t(z)\}_{t \geq 0}$ on D, setting $v(t,z) = h\,(F_t(z))$, we have that $v(t,z) \in \Omega$ for all $t \geq 0$. That is for a fixed $z \in D$, the family $\{v(t,z),\ t \geq 0\}$ determines a continuous curve in Ω, such that $v(0,z) = w$. In addition, it follows by the chain rule that

$$
\frac{\partial v(t,z)}{\partial t} = h'\,(F_t(z))\,\frac{\partial F_t(z)}{\partial t} = -h'\,(F_t(z))\,f\,(F_t(z))\,.
$$

On account of formula (5.2.6) we obtain that

$$
\frac{\partial v(t,z)}{\partial t} = -\mu h(F_t(z)) = -\mu v(t,z).
\tag{5.2.9}
$$

Integrating the latter relation with the initial data

$$
v\,(0,z) = w
\tag{5.2.10}
$$

we obtain

$$
v(t,z) = e^{-\mu t}w \in \Omega \quad \text{for all } t \geq 0.
$$

Since $w \in \Omega$ was arbitrary the first assertion of our proposition follows.

Assume now, in addition, that h has a null point $\tau \in D$ and $h'(\tau) \neq 0$. This implies that there is a neighborhood $U \subset D$ of the point τ such that h is univalent on U and $V = h(U) \subset \Omega$ is a neighborhood of the origin. We want to show that, in fact, h is univalent on the whole of D. Indeed, assuming the contrary, suppose that for some $w \in \Omega$ there are two distinct points z_1 and z_2 in D such that $h(z_1) = h(z_2) = w$.

Observe also that, as a result of equation (5.2.6), the conditions $h(\tau) = 0$ and $h'(\tau) \neq 0$ imply that $f(\tau) = 0$ and $f'(\tau) = \mu$ with $\operatorname{Re} \mu > 0$. Since D is bounded the family $\{F_t(\cdot)\}_{t \geq 0}$ is a normal family in D. Then exactly as in Proposition 4.4.1 one can conclude that $\tau \in D$ is an attractive stationary point of the flow $S_f = \{F_t(\cdot)\}_{t \geq 0}$ generated by f. In particular, we have that $F_t(z_1)$ and $F_t(z_2)$ converge to τ as $t \to \infty$. Note that $F_t(z_1)$ and $F_t(z_2)$ are different for all $t \geq 0$.

As above, now define $v(t, z) = h(F_t(z))$, $t \geq 0$, $z \in D$ and choose $t_0 > 0$ such that $F_t(z_1)$ and $F_t(z_2)$ belong to U for all $t \geq t_0$. Then for such t the curves $v(t, z_i)$, $i = 1, 2$, lie in $V = h(U) \subset \Omega$. But $v(0, z_1) = v(0, z_2) = w$ and we have that $v(t, z_i)$, $i = 1, 2$, are the same as the solutions of the differential equation (5.2.9) with the same initial data (5.2.10). Consequently $F_t(z_1) = F_t(z_2)$ for all $t \geq t_0$. That is a contradiction. $\square$

Sometimes we will say that $h \in \operatorname{Univ}(D, \mathbb{C})$ is μ-spirallike if it satisfies equation (5.2.6) with $\mu \in \mathbb{C}$, $\operatorname{Re} \mu > 0$, and $f \in \mathcal{G}\operatorname{Hol}(D)$.

Remark 5.2.1. Thus $h \in \operatorname{Spiral}(D)$ *is spirallike with respect to an interior point (that is $h \in \operatorname{Spiral}(D, \tau)$ for some $\tau \in D$) if and only if the generator f in (5.2.6) vanishes at $\tau \in D$. Moreover, if f is defined, then $\mu = f'(\tau)$, and τ is an attractive stationary point of the flow* $S_f = \{F_t(\cdot)\}_{t \geq 0}$ *generated by f.* In fact, it can be shown (see Section 5.7) that for each $f \in \mathcal{G}\operatorname{Hol}(D)$, normalized by the conditions $f(\tau) = 0$ and $\operatorname{Re} f'(\tau) > 0$ there is a unique solution $h \in \operatorname{Spiral}(D, \tau)$ of the equation (5.2.6) with $\mu = f'(\tau)$ normalized by $h'(\tau) = 1$. In addition, *if μ is a purely real number, then h defined by (5.2.6) is a starlike function on D.*

Similarly, we can say that $h \in \operatorname{Snail}(D)$ *if and only if f has no null point in* D. In this case the flow $S_f = \{F_t(\cdot)\}_{t \geq 0}$ generated by f for each $z \in D$ converges to a boundary point $\tau \in \partial D$.

Since for the special (but most important) case when $D = \Delta$ is the unit disk we have a complete description of the class $\mathcal{G}\operatorname{Hol}(\Delta)$, the proved propositions imply several corollaries. In particular, applying the Berkson–Porta parametric representation of the class $\mathcal{G}\operatorname{Hol}(\Delta)$ we obtain the following:

Corollary 5.2.1 *Let $h : \Delta \mapsto \mathbb{C}$ be a univalent holomorphic function on Δ. Then $h(\Delta)$ is spiralshaped if and only if the following equation is fulfilled:*

$$\mu h(z) = h'(z)(z - \tau)(1 - z\bar{\tau})p(z), \quad z \in \Delta, \tag{5.2.11}$$

where $\tau \in \overline{\Delta}$, $\mu \in \mathbb{C}$ with $\operatorname{Re} \mu > 0$, and $p \in \operatorname{Hol}(\Delta, \mathbb{C})$ with $\operatorname{Re} p(z) \geq 0$ for all $z \in \Delta$.

Remark 5.2.2 Thus, if $\tau \in \Delta$, then $h \in \operatorname{Spiral}(\Delta, \tau)$ (i.e., is spirallike with respect to an interior point); if $\tau \in \partial\Delta$, then $h \in \operatorname{Snail}(\Delta)$ (i.e., spirallike with respect to a boundary point). Separating these two cases we conclude:

A locally biholomorphic function h on Δ belongs to $\operatorname{Spiral}(\Delta, \tau)$ if and only if there exist $\tau \in \Delta$ and $\mu \in \mathbb{C}$ with $\operatorname{Re} \mu > 0$ such that

$$\operatorname{Re} \frac{h'(z)(z - \tau)(1 - z\bar{\tau})}{\mu h(z)} > 0, \quad z \in \Delta.$$

Indeed, equation (5.2.11) can be rewritten in the form

$$\text{Re}\,\frac{h'(z)(z-\tau)(1-z\bar{\tau})}{\mu h(z)} = \text{Re}\,\frac{1}{p(z)} \geq 0, \quad z \in \Delta. \qquad (5.2.11')$$

If $\tau \in \Delta$, then differentiating (5.2.11) at $z = \tau$ we obtain $p(\tau) = \dfrac{\mu}{1-|\tau|^2}$, which means that inequality (5.2.11') is actually strict.

If we normalize μ by the condition $|\mu| = 1$, we obtain that τ and p are uniquely determined by h. Of course, the converse assertion is also true. Moreover, if $\tau = 0$ then setting $\theta = -\arg\mu$ we obtain a description of the set Spiral$(\Delta, 0)$ which coincides with the classical description (Proposition 5.2.3).

Letting $\mu = 1$ and $\tau = 0$ in (5.2.11') we arrive immediately at Nevanlinna's condition (Proposition 5.2.1) and Wald's condition if $\tau \in \Delta$, $\tau \neq 0$ (Proposition 5.2.2).

Similarly, we conclude from (5.2.11):

A univalent function h on Δ belongs to Snail(Δ) *(respectively,* Fan(Δ)*) if and only if for some $\tau \in \partial\Delta$ and $\mu \in \mathbb{C}$ (respectively, $\mu \in \mathbb{R}$) with* Re $\mu \geq 0$, *the following condition holds:*

$$\text{Re}\,\frac{h'(z)(z-\tau)^2}{\mu h(z)\tau} \leq 0.$$

We will see in the next section that the boundary behavior (at the point τ) of the quotient $Q(z) = \dfrac{h'(z)(z-\tau)}{h(z)}$ (the so called Visser–Ostrowski quotient) in the latter inequality characterizes those functions in Snail(Δ) which, in fact, belong to Fan(Δ).

5.3 A generalized Visser–Ostrowski condition and fanlike functions

In this section some relations between classes Snail(Δ) and Fan(Δ) (of spirallike and starlike functions with respect to a boundary point) will be studied.

First we make the following observation.

Suppose that $h \in$ Snail(Δ), that is $h \in$ Univ(Δ) satisfies equation (5.2.6):

$$\mu h(z) = h'(z)f(z), \quad z \in \Delta,$$

for some $\mu \in \mathbb{C}$ with Re $\mu > 0$, and $f \in \mathcal{G}\,\text{Hol}(\Delta)$ with no null point in Δ.

We know that in this case there is a unique boundary point $\tau \in \partial\Delta$ which is the sink point of the semigroup generated by f.

Assume, temporarily, that one of the following conditions holds:

(i) $|\dfrac{h''(z)}{h'(z)}| \leq M < \infty, \quad z \in \Delta,$

or

(ii) $\lim\limits_{z \to \tau} h(z) = 0$ and $\lim\limits_{z \to \tau} h'(z) = a, \quad a \neq 0, \infty,$ as λ approaches τ nontangentially.

The latter condition is known to define h to be *conformal at the boundary point* $\tau \in \partial\Delta$ (see, for example, [107]).

Without loss of generality one can set $\tau = 1$. We recall, that in our situation as a result of the Berkson–Porta formula $f \in \mathcal{G}\operatorname{Hol}(\Delta)$ has the form:

$$f(z) = -q(z)(1-z)^2, \tag{5.3.1}$$

with some $q \in \operatorname{Hol}(\Delta, \mathbb{C})$ such that $\operatorname{Re} q \geq 0$ everywhere.

In addition, we know by the established continuous version of the Julia–Wolff–Carathéodory Theorem (Proposition 4.6.1) that the angular derivative

$$\angle \lim_{z \to 1} \frac{f(z)}{z-1}$$

of f exists at 1 and equals to the angular limit

$$\angle \lim_{z \to 1} f'(z) = \beta \geq 0. \tag{5.3.2}$$

We want to show that in both cases (i) and (ii) the number μ in equation (5.2.6) must be equal to β, that is, μ is actually a real number. Indeed, assuming first that (i) holds and differentiating equation (5.2.6), we obtain

$$\mu - f'(z) = \frac{h''(z)}{h'(z)} f(z). \tag{5.3.3}$$

Since $\lim\limits_{r \to 1^-} f(r) = 0$ we obtain $\mu = \beta$ by (5.3.2).

If now condition (ii) holds, then one can write, by using equalities (5.2.6) and (5.3.1), that $\mu h(z) = -h'(z)(1-z)^2 q(z)$ with $\operatorname{Re} q(z) \geq 0$, $z \in \Delta$. Then we have

$$\mu \frac{1}{z-1} h(z) = h'(z)(1-z)q(z) = h'(z)\frac{f(z)}{z-1}. \tag{5.3.4}$$

Since

$$\angle \lim_{z \to 1} \frac{h(z)}{z-1} = \angle \lim_{z \to 1} h'(z) = a \neq 0, \infty,$$

we see again that $\mu = \beta$ is real, hence h is, actually, a fanlike function. $\quad\square$

Of course, these simple considerations are related to the question of how can one join conditions (i) and (ii).

For example, one can replace condition (ii) by the condition that h be *isogonal* at τ, that is h and $\arg h'$ have finite angular limits at τ. It is clear that if h is conformal at τ then it is isogonal at τ. In general the converse does not hold. At the same time some geometrical arguments indicate that there is no properly spirallike function with respect to a boundary point which is isogonal at it.

Another (much weaker than isogonality) condition which is sufficient to the above mentioned property is the Visser–Ostrowski condition:

$$\angle \lim_{z \to \tau} \frac{h'(z)(z - \tau)}{h(z) - h(\tau)} = 1. \tag{5.3.5}$$

It is known (see [107]) that this condition is equivalent to the following:

$$\angle \lim_{z \to \tau} (z - \tau) \frac{h''(z)}{h'(z)} = 0. \tag{5.3.6}$$

Obviously condition (i) implies (5.3.6). Thus the Visser–Ostrowski condition is weaker than both (i) and (ii). We will establish now that a generalized Visser–Ostrowski condition (see condition (*) below) is necessary and sufficient for $h \in$ Snail(Δ) to be in Fan(Δ).

Let us consider the Visser–Ostrowski quotient:

$$Q(z) = \frac{h'(z)(z - \tau)}{h(z) - h(\tau)}. \tag{5.3.7}$$

Proposition 5.3.1 ([38]) *Let h be a univalent function on Δ such that $h(\Delta)$ is a spiralshaped set with*

$$\lim_{r \to 1^-} h(r\tau) = 0, \quad \tau \in \partial\Delta. \tag{5.3.8}$$

Then $h(\Delta)$ is, in fact, starshaped if and only if the following condition holds:

(*)   *the angular limit* $\angle \lim_{z \to \tau} Q(z) := \nu$

exists finitely and is a positive real number.

Proof. Assuming again that the generator f in equation (5.2.6) is presented by the Berkson–Porta formula (5.3.1), we could use a similar argument as in (5.3.4) if we show that the point $\tau \in \partial\Delta$ in (5.3.7), (5.3.8) and (*) is equal to 1.

Indeed, if this is not the case we have by (5.2.6) and (5.3.1) that

$$\frac{\mu h(z)}{h'(z)(z - \tau)} = -\frac{(1 - z)^2 q(z)}{(z - \tau)}. \tag{5.3.9}$$

Hence by (*) the limit

$$\eta := \angle \lim_{z \to \tau} \left[-(1 - z)^2 \frac{q(z)}{z - \tau} \right] = \frac{\mu}{\nu} \tag{5.3.10}$$

exists (finitely). On the other hand, setting $z = r\tau$, $r \in (0, 1)$, and letting r approach 1^- we obtain

$$\eta = (1 - \tau)(\tau - 1)\bar{\tau} \lim_{r \to 1^-} \frac{q(r\tau)}{r - 1} = |1 - \tau|^2 \lim_{r \to 1^-} \frac{q(r\tau)}{r - 1}$$

and

$$\operatorname{Re}\eta = |1 - \tau|^2 \lim_{r \to 1^-} \operatorname{Re}\frac{q(r\tau)}{r-1} \le 0.$$

But the latter equality is impossible because of (5.3.10) and $\operatorname{Re}\mu > 0$. Contradiction. So $\tau = 1$.

Once again, since

$$\mu h(z) = \frac{f(z)}{(z-1)}h'(z)(z-1)$$

we obtain

$$\mu = \angle \lim_{z \to 1}\frac{f(z)}{(z-1)}Q(z) = \beta\angle\lim_{z \to 1}Q(z),$$

and our assertion follows. $\square$

Thus the class of properly spirallike (i.e., not starlike) functions with respect to a boundary point contains neither conformal nor isogonal mappings at this boundary point.

Remark.5.3.1. We will show below (see Section 5.6) that the number $\nu = \angle\lim_{z \to \tau}Q(z)$ yields an important geometrical characteristic of $h \in \operatorname{Fan}(\Delta)$, namely, the angle $\theta = \pi\nu$ is the smallest one such that $h(\Delta)$ lies in the wedge of the angle θ. Thus, in fact, for $h \in \operatorname{Fan}(\Delta)$ with $\lim_{r \to 1^-}h(r) = 0$ we will show that
$\angle\lim_{z \to \tau}Q(z) \le 2$.

To do this we need an approximation process which is based on Hummel's representation of the class $\operatorname{Star}(\Delta, \cdot)$ of starlike functions with respect to interior points. We will give this representation in Section 5.5.

5.4   An invariance property and approximation problems

The following question naturally arises in approximation function theory: given $h \in \operatorname{Hol}(\Delta, \mathbb{C})$ such that $\Omega = h(\Delta)$ is spiralshaped (respectively, starshaped) and a sequence $\Omega_n \subset \Omega$ spiralshaped (starshaped) domains such that $\bigcup\Omega_n = \Omega$, find the sequence h_n such that $\lim_{n \to \infty}h_n(z) = h(z)$ for each $z \in \Delta$, and $h_n(\Delta) = \Omega_n$.

Theoretically, under certain conditions this problem can be solved because of the Riemann Mapping Theorem and Carathéodory's Kernel Convergence Theorem.

However, in general, to find such a sequence implicitly has not seemed plausible, even the sequence Ω_n is well described.

At the same time one can define, in a sense, a dual approximation problem which is related to some invariance property of spirallike (respectively, starlike) functions.

Namely, if $h \in \text{Spiral}(\Delta)$ (respectively, $\text{Star}(\Delta)$), find a 'nice' sequence of domains D_n in Δ such that $\bigcup D_n = \Delta$ and h continue to be spirallike (respectively, starlike) on each D_n.

In view of Propositions 5.2.4 and 5.2.5 an answer to the above question is provided by the following observation. If $h \in \text{Spiral}(\Delta)$ (respectively, $\text{Star}(\Delta)$), then for some $\mu \in \mathbb{C}$ (respectively, $\mu \in \mathbb{R}$) with $\text{Re}\,\mu > 0$ the function $f(z) = \mu h'(z)[h(z)]^{-1}$ is a semi-complete vector field on Δ. Then these propositions yield that h continues to be spirallike (respectively, starlike) on a domain $D \subset \Delta$ if and only if this domain is invariant for the semigroup $S = \{F_t\}_{t \geq 0}$ generated by f.

Thus, due to Proposition 5.2.4 we can formulate the following assertion:

Proposition 5.4.1 *Let $h \in \text{Spiral}(\Delta)$ (respectively, $h \in \text{Star}(\Delta)$). Then there is a unique point $\tau \in \overline{\Delta}$ such that for all $K > 1 - |\tau|^2$ the sets $h(D(\tau, K))$, where*

$$D(\tau, K) = \left\{ z \in \Delta : \frac{|1 - z\bar{\tau}|^2}{1 - |z|^2} < K \right\},$$

are spiralshaped (respectively, starshaped).

We recall that for $\tau \in \Delta$ the sets $D(\tau, K) = \{ z \in \Delta : |\frac{z - \tau}{1 - \bar{\tau}z}| < r \}$, $K = (1 - |\tau|^2)(1 - r^2)^{-1}$ (see Section 1.1, Exercise 6) are pseudo-hyperbolic balls in Δ while for $\tau \in \partial\Delta$ these sets are horocycles internally tangent to $\partial\Delta$ at τ.

In particular, for $\tau = 0$ we have that $h \in \text{Spiral}(\Delta, 0)$ (respectively, $\text{Star}(\Delta, 0)$) is spirallike (respectively, starlike) on each disk $\Delta_r = \{|z| < r, \ 0 < r \leq 1\}$ concentric with Δ.

(Note that for the functions of the class

$$S^* = \{ h \in \text{Star}(\Delta) : h(0) = 0 \text{ and } h'(0) = 1 \}$$

this result was obtained independently by Takahashi and Seidel [18] as an extension to Nevanlinna's theorem (see Proposition 5.2.1). A simple proof of this fact with the use of the Schwarz Lemma can be found in [33]).

One may expect that even when $\tau \neq 0$ and $h \in \text{Spiral}(\Delta, \tau)$ (respectively, $h \in \text{Star}(\Delta)$) on the unit disk, then for at least r close enough to 1 it continues to be spirallike (starlike) on the disks Δ_r. However, examples show that in general this conjecture has been answered in the negative.

Example 1. Consider the function $h(z) = h_0(M_{\frac{1}{2}}(z))$, $z \in \Delta$, where

$$h_0(z) = \frac{z}{(1 - z)^2}$$

is the so called Koebe function and

$$M_{1/2}(z) = \frac{\frac{1}{2} - z}{1 - \frac{1}{2}z}$$

is a Möbius involution transformation of the unit disk taking the origin to $\tau = \frac{1}{2}$. It is easy to see that the Koebe function belongs to $\text{Star}(\Delta, 0)$ (even S^*) and it follows that $h \in \text{Star}(\Delta, \tau)$, because of the relations $h(\Delta) = h_0(\Delta)$, $h(\tau) = 0$. We claim that there is a sequence $\{z_n\} \subset \Delta$, $|z_n| \to 1$ such that

$$\text{Re} \left[\frac{zh'(z_n)}{h(z_n)} \right] < 0.$$

Indeed, calculating

$$\text{Re} \left[\frac{zh'(z)}{h(z)} \right] = \frac{9}{4} \frac{\text{Re} \, z(1-z)(1+\bar{z})(\bar{z} - \frac{1}{2})(1 - \frac{1}{2}\bar{z})}{|1+z|^2 |z - \frac{1}{2}|^2 |1 - \frac{1}{2}z|^2}$$

and setting $z = x + iy$ we obtain (after several technical manipulations) that the numerator $N(z)$ of the right hand side of the latter equation has the form

$$N(z) = \frac{1}{2}(x - \frac{1}{2})(x^2 + y^2 - 1)(x^2 + y^2 - 2x).$$

Now it can easily be seen that for any sequence $\{z_n\} \in \Delta$, $|z_n| \to 1$ such that $z_n \in D = \{z = x + iy : (x - \frac{1}{2})(x^2 + y^2 - 2x) > 0\}$ the expression $N(z)$ is negative (see Figure 5.4). Thus the claimed assertion is proved.

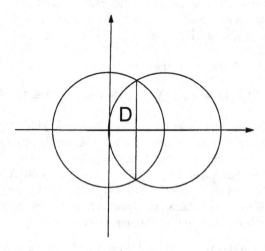

Figure 5.4: The set D.

Nevertheless, it has been shown in [66] that if τ is close enough to zero, then the answer to the above question (*concerning starlike functions*) is affirmative.

Proposition 5.4.2 *Let $h \in \text{Star}(\Delta, \tau)$ with $|\tau| < 2 - \sqrt{3}$. Then there exists $\varepsilon > 0$ such that for each $1 - \varepsilon < r \leq 1$ the function h belongs to $\text{Star}(\Delta_r)$.*

By using the results in Sections 3.6 and 4.5 we can formulate some sufficient conditions for $h \in \text{Spiral}(\Delta, \tau)$ (respectively, $\text{Star}(\Delta, \tau)$) to be spirallike (respectively, starlike) on each Δ_r for r close enough to 1, which do not depend on the location of $\tau \in \Delta$.

We will say that a function $h \in \text{Hol}(\Delta, \mathbb{C})$ is strongly spirallike if for some $\mu \in \mathbb{C}$ with $\text{Re}\,\mu > 0$ the function $f(z) := \dfrac{\mu h(z)\bar{z}}{h'(z)}$ is strongly ρ-monotone on Δ.

Proposition 5.4.3 *Let $h \in \text{Hol}(\Delta, \mathbb{C})$ be a strongly spirallike function. Then there is $\tau \in \Delta$ and $\varepsilon > 0$ such that for each $1 - \varepsilon < r \leq 1$, $h \in \text{Spiral}(\Delta_r)$ and $h(\tau) = 0$.*

In particular, by using a criterion of strong ρ-monotonicity we have the following:

Corollary 5.4.1 *Let α be a real continuous function on $[0, 1]$ such that $\alpha(1) > 0$ and let $h \in \text{Hol}(\Delta, \mathbb{C})$ be such that for some $\mu \in \mathbb{C}$ with $\text{Re}\,\mu > 0$ the following condition holds*

$$\text{Re}\,\frac{\mu h(z)\bar{z}}{h'(z)} \geq \alpha(|z|)|z|, \qquad z \in \Delta.$$

Then there exists $\tau \in \Delta$ and $\varepsilon > 0$ such that for each $1 - \varepsilon < r \leq 1$

$$h \in \text{Spiral}(\Delta_r) \text{ and } h(\tau) = 0.$$

Exercise 1. Let $f(z) = az^2 - a + bz$ for some positive a and b.

(i) Show that if equation (5.2.6) is solvable for some $\mu \in \mathbb{C}$, $\text{Re}\,\mu > 0$, then μ is, in fact, a positive real number, that is, the solution h of (5.2.6) is, actually, a starlike function on Δ. Find μ.

(ii) Show that there exists $\varepsilon > 0$ such that for each $r \in (1-\varepsilon, 1]$, $h \in \text{Star}(\Delta_r)$.

(iii) Set $a = 2$ and $b = 3$. Show that ε in (ii) can be chosen arbitrarily in the interval $[0, \frac{1}{2})$.

Exercise 2. Let $f(z) = az^2 - \bar{a} + bz\dfrac{1 - cz}{1 + cz}$ with some complex numbers a, b and c such that $\text{Re}\,b > 0$ and $|c| < 1$.

(i) Show that there are $\mu \in \mathbb{C}$ and $\varepsilon > 0$ such that equation (5.2.6) has a solution $h \in \text{Spiral}(\Delta_r, \tau)$ for some $\tau \in \Delta$ and each $r \in (1 - \varepsilon, 1]$.

(ii) Find relations between a, b and c such that μ is real, i.e., h is starlike.

(iii) Setting $a = 0$ and $b > 0$, show that h defined by (5.2.6) with an appropriate $\mu > 0$ is starlike of order α (see Section 5.2) with $\alpha = b\dfrac{1 - |c|}{1 + |c|}$.

Finally, note that if $h \in \mathrm{Hol}(\Delta, \mathbb{C})$ is known to be in $\mathrm{Spiral}(\Delta)$, then sometimes information on values of h and its first and second derivatives at only one point may be useful for solving the dual approximation problem on disks Δ_r concentric with Δ.

Indeed, using Corollary 4.5.1 and Remark 4.5.2 we can easily arrive at the following sufficient condition:

Proposition 5.4.4 *Let* $h \in \mathrm{Spiral}(\Delta)$ *(respectively,* $\mathrm{Star}(\Delta)$*), satisfy equation (5.3.6) with some* $\mu \in \mathbb{C}$ *(respectively,* $\mu \in \mathbb{R}$*) with* $\mathrm{Re}\,\mu > 0$ *and* $f \in \mathrm{Hol}(\Delta, \mathbb{C})$*. Then*

(i) $\mathrm{Re}\,\mu \dfrac{h''(0)h(0)}{[h'(0)]^2} < \mathrm{Re}\,\mu;$

(ii) if $4\left|\dfrac{h(0)}{h'(0)}\right| < \left[1 - \dfrac{h''(0)h(0)}{[h'(0)]^2}\right]$ *then for some* $\varepsilon > 0$ *and each* $r \in (1 - \varepsilon, 1]$
the function $h \in \mathrm{Spiral}(\Delta_r)$ *(respectively,* $\mathrm{Star}(\Delta_r)$*).*

Remark 5.4.1. Of course, the problem above is not relevant for the class $\mathrm{Fan}(\Delta)$. Indeed, it is clear that if $h \in \mathrm{Star}(\Delta)$ *has no null points in* Δ then there is no disk Δ_r concentric with Δ such that $h(\Delta_r)$ is starshaped.

In this case another approximation problem arises: *given* $h \in \mathrm{Fan}(\Delta)$*; find a sequence* $h_n \in \mathrm{Star}(\Delta, \cdot)$ *of starlike functions with interior null points such that* h_n *converge to* h *uniformly on compact subsets of* Δ *as* $n \to \infty$.

It looks as if the following procedure should work in solving this problem. If, for example, $h \in \mathrm{Fan}(\Delta)$ is isogonal at its boundary null point, say $\tau = 1$, then it satisfies the equation

$$\mu h(z) = h'(z) f(z), \quad z \in \Delta, \tag{5.4.1}$$

for some $f \in \mathcal{G}\,\mathrm{Hol}(\Delta)$:

$$f(z) = -q(z)(1 - z)^2,$$

with $\mathrm{Re}\,q(z) > 0$, $z \in \Delta$, and $\mu = \angle f'(1) > 0$. Define an approximation sequence f_n to f by using the Berkson–Porta representation formula $f_n(z) = (z - \tau_n)(1 - \bar{\tau}_n z)q(z)$, where $\{\tau_n\} \subset \Delta$ is any sequence which converges to 1 as n tends to ∞. If, in addition, we can choose this sequence such that the numbers $f_n'(\tau_n)$ are real, then we may try to solve the equations

$$\mu_n h_n(z) = h_n'(z) f_n(z), \quad z \in \Delta, \tag{5.4.2}$$

with $\mu_n = f'(\tau_n)$ in order to define a sequence of univalent functions h_n which are starlike with respect to interior points $(h(\tau_n) = 0)$. However, in this way there is a risk that h_n may not converge to h. Indeed, putting it otherwise, we obtain that the numbers $\mu_n = f_n'(\tau_n) = (1 - |\tau_n|^2)q(\tau_n)$ must converge to

$$\mu = \angle \lim_{z \to 1} f'(z) = \angle \lim_{z \to 1} \frac{f(z)}{z - 1} = \angle \lim_{z \to 1} (1 - z)q(z).$$

On the other hand, if τ_n achieves $\tau = 1$ along the real axis we have

$$\lim_{n \to \infty} \mu_n = \lim_{n \to \infty} (1 - |\tau_n|^2)q(\tau_n) = 2 \lim_{n \to \infty} (1 - \tau_n)q(\tau_n) = 2\mu.$$

That is a contradiction.

Also it is easy to construct a counter example of the latter relation.

Example 2. Consider a semi-complete (even complete) vector field $f \in$ Hol$(\Delta, \mathbb{C})$ defined as follows:

$$f(z) = z^2 - 1.$$

There are two boundary null points $z = 1$ and $z = -1$ of f. Since $f'(1) = 2 > 0$ it follows by Proposition 4.4.1 (a continuous version of the Julia–Wolff–Carathéodory Theorem) that $z = 1$ is a sink point of the semigroup generated by f. Therefore, one can present f by the Berkson–Porta formula:

$$f(z) = -(1-z)^2 \frac{1+z}{1-z}$$

with $q(z) = \dfrac{1+z}{1-z}$. Obviously, q admits real values if and only if $z \in \Delta$ are real. So, we have to choose a sequence $\{\tau_n\} \in \Delta$, $\tau_n \to 1$, in the above approximation procedure to be real. Further, if we define

$$f_n(z) = (z - \tau_n)(1 - \overline{\tau_n} z) \frac{1+z}{1-z},$$

we obtain $f_n(\tau_n) = 0$, and $f_n(z) \to f(z)$ in Δ, while

$$f_n'(\tau_n) = (1 - \tau_n^2) \frac{1 + \tau_n}{1 - \tau_n} = (1 + \tau_n)^2 \to 4 \neq f'(1).$$

As a matter of fact, it can be shown that if h_n are solutions of (5.4.2) normalized by the condition $h_n(0) = 1$, then they converge to a fanlike function $\hat{h}$ which is a power of h defined by (5.2.1). However, in general, we do not know whether the numbers μ_n are real (i.e., whether the functions h_n are starlike).

Thus a procedure of using the Berkson–Porta multiplier $(z - \tau_n)(1 - \overline{\tau_n} z)$ has been shown to be inappropriate for the approximation of starlike functions with respect to a boundary point by starlike functions with respect to interior points. Nevertheless, it turns out that a modification of this multiplier in the spirit of J.A. Hummel makes such a procedure very effective. Hummel's multiplier has been used by A. Lyzzaik to prove a conjecture of M.S. Robertson on a description of starlike functions with respect to a boundary point, the images of which lie in a half plane. The next section is devoted to this approach.

5.5   Hummel's multiplier and parametric representations of starlike functions

Hummel's multiplier is a (meromorphic) function of the form

$$H_\tau(z) = \frac{(z - \tau)(1 - z\bar{\tau})}{z}, \qquad z \in \Delta,$$

where τ is a given complex number with $|\tau| \leq 1$.

For the case of $|\tau| < 1$, J.A. Hummel showed in [65] and [67] that this function plays a special role in the study of starlike functions. In fact, it turns out that by the multiplication operation this function translates the set $\mathrm{Star}(\Delta, 0)$ onto the set $\mathrm{Star}(\Delta)$. Moreover, if $\tau \in \Delta$ then $\mathrm{Star}(\Delta, 0)$ is translated onto $\mathrm{Star}(\Delta, \tau)$. More precisely:

Proposition 5.5.1 *Let* $h \in \mathrm{Hol}(\Delta, \mathbb{C})$, $h(0) = 0$, *and* $g \in \mathrm{Hol}(\Delta, \mathbb{C})$ *be two functions related by the formula*

$$g(z) = H_\tau(z) \cdot h(z). \tag{5.5.1}$$

Then $g(\Delta)$ *is starshaped if and only* $h(\Delta)$ *is starshaped.*

In this section we will treat only the case when $\tau \in \Delta$. In this situation it is more convenient (for some symmetry) to consider the meromorphic function Ψ_τ on Δ defined by

$$\Psi_\tau(z) = \frac{1}{1 - |\tau|^2} H_\tau(z) = \frac{1}{1 - |\tau|^2} \cdot \frac{(z - \tau)(1 - z\bar{\tau})}{z}, \qquad z \in \Delta. \tag{5.5.2}$$

Of course, it is sufficient to prove Proposition 5.5.1 replacing H_τ in (5.5.2) by Ψ_τ. Moreover, in this case the inverse translation is just the same multiplier composed with a Möbius involution transformation.

Proposition 5.5.2 *Let* $h \in \mathrm{Hol}(\Delta, \mathbb{C})$ *and* $g \in \mathrm{Hol}(\Delta, \mathbb{C})$ *be two functions related by formula:*

$$g(z) = \Psi_\tau(z) \cdot h(z). \tag{5.5.3}$$

Then $g \in \mathrm{Star}(\Delta, \tau)$ *if and only if* $h \in \mathrm{Star}(\Delta, 0)$. *Moreover, the inverse transformation can be defined by the formula*

$$h(z) = \Psi_\tau(M_\tau(z)) \cdot g(z),$$

where

$$M_\tau(z) = \frac{\tau - z}{1 - z\bar{\tau}}.$$

To prove this assertion we need some properties of the function Ψ_τ defined by (5.5.2).

Lemma 5.5.1 *For the function* Ψ_τ *the following properties hold:*

(i) for all $z \in \partial\Delta : |z| = 1$, $\Psi_\tau(z)$ *is a positive real number;*

(ii) for all $z \in \partial\Delta : |z| = 1$,

$$\operatorname{Re} \frac{z\Psi'_\tau(z)}{\Psi_\tau(z)} = 0;$$

(iii) if M_τ *denotes the Möbius transformation*

$$M_\tau(z) = \frac{\tau - z}{1 - z\bar{\tau}},$$

then

$$\Psi_\tau(z) \cdot \Psi_\tau(M_\tau(z)) = 1.$$

Proof.

(i) If we set $z = e^{i\varphi}$, $0 \leq \varphi \leq 2\pi$ we obtain:

$$\begin{aligned}
\Psi_\tau(e^{i\varphi}) &= \frac{(e^{i\varphi} - \tau)(1 - e^{i\varphi}\bar{\tau})}{(1 - |\tau|^2)e^{i\varphi}} = \frac{(1 - e^{i\varphi}\tau)(1 - e^{i\varphi}\bar{\tau})}{1 - |\tau|^2} \\
&= \frac{|1 - e^{i\varphi}\bar{\tau}|^2}{1 - |\tau|^2} > 0.
\end{aligned}$$

(ii) By direct calculations we have:

$$\begin{aligned}
\Psi'_\tau(z) &= \frac{1}{1 - |\tau|^2} \left[\left(1 - \frac{\tau}{z}\right)(1 - z\bar{\tau}) \right] \\
&= \frac{1}{1 - |\tau|^2} \left[\frac{\tau}{z^2}(1 - z\bar{\tau}) - \left(1 - \frac{\tau}{z}\right)\bar{\tau} \right] \\
&= \frac{1}{1 - |\tau|^2} \left[\frac{\tau}{z^2} - \frac{|\tau|^2}{z} - \bar{\tau} + \frac{|\tau|^2}{z} \right] = \frac{1}{1 - |\tau|^2} \left(\frac{\tau}{z^2} - \bar{\tau} \right).
\end{aligned}$$

Again setting $z = e^{i\varphi}$ we obtain:

$$\begin{aligned}
\operatorname{Re} \frac{z\Psi'_\tau(z)}{\Psi_\tau(z)} &= \operatorname{Re} \frac{e^{2i\varphi}\left(\frac{\tau}{e^{2i\varphi}} - \bar{\tau}\right)}{(e^{i\varphi} - \tau)(1 - e^{i\varphi}\bar{\tau})} = \operatorname{Re} \frac{\tau - \bar{\tau}e^{2i\varphi}}{e^{i\varphi}|1 - e^{-i\varphi}\tau|^2} \\
&= \frac{1}{|1 - e^{-i\varphi}\tau|^2} \cdot \operatorname{Re}(\tau e^{-i\varphi} - \bar{\tau}e^{i\varphi}) = 0.
\end{aligned}$$

(iii) Substitute $M_\tau(z) = \dfrac{\tau - z}{1 - z\bar{\tau}}$ to Ψ_τ instead of z. We obtain:

$$\begin{aligned}
\Psi_\tau(M_\tau(z)) &= \frac{(1 - z\bar{\tau})}{1 - |\tau|^2} \cdot \frac{\left(\frac{\tau - z}{1 - z\bar{\tau}} - \tau\right)\left(1 - \frac{\tau - z}{1 - z\bar{\tau}} - \bar{\tau}\right)}{\tau - z} \\
&= \frac{(\tau - z - \tau + z|\tau|^2)(1 - z\bar{\tau} + z\bar{\tau} - |\tau|^2)}{(1 - |\tau|^2)(\tau - z)(1 - z\bar{\tau})} \\
&= \frac{-z(1 - |\tau|^2)}{(\tau - z)(1 - z\bar{\tau})} = \frac{1}{\Psi_\tau(z)}.
\end{aligned}$$

The lemma is proved. □

Following J.A. Hummel [65, 67] we will say that *a meromorphic function g on* Δ *satisfies the condition (A) if for each* $\varepsilon > 0$ *there is* $\rho \in (0,1)$ *such that for all* z *in the annulus* $\rho < |z| < 1$ *the following condition holds:*

$$\text{Re } \frac{zg'(z)}{g(z)} > -\varepsilon. \tag{5.5.4}$$

Lemma 5.5.2 *Let* $g \in \text{Star}(\Delta, \tau)$ *for some* $\tau \in \Delta$. *Then* g *satisfies condition (A).*

Proof. Since $g(\Delta)$ is starshaped and the origin is an interior point of $g(\Delta)$, the function g satisfies the equation:

$$g(z) = g'(z)(z - \tau)(1 - z\bar{\tau})p(z)$$

with some $\tau \in \Delta$, such that $g(\tau) = 0$ and $p \in \text{Hol}(\Delta, \mathbb{C})$ with $\text{Re } p > 0$ everywhere. Then by Lemma 5.5.1(iii) we obtain:

$$(1 - |\tau|^2) \cdot \frac{z \cdot g'(z)}{g(z)} = \frac{1}{\Psi_\tau(z) \cdot p(z)} = \frac{\Psi_\tau(M_\tau(z))}{p(z)}. \tag{5.5.5}$$

Denoting $w = M_\tau(z)$ we have in turn from (5.5.4):

$$\frac{z \cdot g'(z)}{g(z)} = \Psi_\tau(w) \cdot q(w), \tag{5.5.6}$$

where $q(w) = \dfrac{1}{(1 - |\tau|^2)p(M_\tau(w))}$ (recall, that M_τ is an involution).

Hence $\text{Re } q(w) > 0$ for all $w \in \Delta$. Now for each $w \in \Delta$ the right hand side of (5.5.5) can be presented as:

$$\Psi_\tau(w) \cdot q(w) = \frac{(w - \tau)(1 - w\bar{\tau})\, q(w)}{(1 - |\tau|^2)w} = \frac{f(w)}{w}, \tag{5.5.7}$$

where $f(z) = \dfrac{(w - \tau)(1 - w\bar{\tau})\, q(w)}{(1 - |\tau|^2)}$ defines a semi-complete vector field by the Berkson–Porta representation.

At the same time, it follows by Proposition 3.4.4 that f satisfies the condition:

$$\text{Re } f(w) \cdot \bar{w} \geq \text{Re } f(0)\bar{w}(1 - |w|^2).$$

Therefore for each $\varepsilon > 0$ one can find $\rho_1 \in (0,1)$ such that:

$$\text{Re } \frac{f(w)}{w} = \frac{1}{|w|^2} \text{Re } f(w)\bar{w} \geq -\varepsilon, \tag{5.5.8}$$

whenever $\rho_1 < |w| < 1$.

Note now, that if w runs in the annulus $\rho_1 < |w| < 1$, then $z = M_\tau(w)$ runs in the set $A = \Delta \setminus B(\tau, r)$ where $B(\tau, r) \subset \Delta$ is a hyperbolic ball centered at τ.

But this ball is also strictly inside Δ, hence there is $\rho \in (0,1)$ such that the annulus $\rho < |z| < 1$ lies in A (see Figure 5.5).

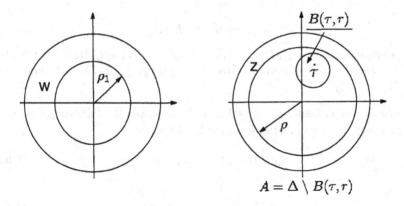

Figure 5.5: Condition (A).

Thus for such $z : \rho < |z| < 1$ we obtain from (5.5.5)–(5.5.7) relation (5.5.3). The lemma is proved. $\square$

Proof of Proposition 5.5.2. Let $h \in \text{Hol}(\Delta, \mathbb{C})$, and $g \in \text{Hol}(\Delta, \mathbb{C})$ be related by (5.5.2). Assume that $g \in \text{Star}(\Delta, \tau)$. It is clear that τ must be a unique null point of g.

Note, that Ψ_τ satisfies condition (A) due to Lemma 5.5.1(ii).

By differentiating (5.5.2) we obtain the equality:

$$\frac{z \cdot h'(z)}{h(z)} = \frac{z \cdot g'(z)}{g(z)} - \frac{z\Psi'(z)}{\Psi(z)},$$

which implies that h also satisfies condition (A). At the same time, since $h \in \text{Hol}(\Delta, \mathbb{C})$ we must have $h(0) = 0$ and $h'(0) = -\dfrac{g(0)}{\tau} \neq 0$. Therefore it follows by the minimum principle for harmonic functions that actually

$$\text{Re}\,\frac{zh'(z)}{h(z)} \geq 0 \qquad (5.5.9)$$

for all $z \in \Delta$. Thus $h \in \text{Star}(\Delta, 0)$.

The converse assertion can be easily proved by replacing the roles of h and g.

Indeed, let now $h \in \text{Hol}(\Delta, \mathbb{C})$, $h(0) = 0$ be starlike, and $g \in \text{Hol}(\Delta, \mathbb{C})$ be defined by (5.5.2) with some $\tau \in \Delta$.

Note that a function $g_1 \in \text{Hol}(\Delta, \mathbb{C})$ defined by:

$$g_1(z) = h(M_\tau(z))$$

is also starlike because:

$$g_1(\Delta) = h(\Delta).$$

Again, using Lemma 5.5.1(iii) we obtain:

$$g(M_\tau(z)) = h(M_\tau(z)) \cdot \Psi_\tau(M_\tau(z)) = \frac{g_1(z)}{\Psi_\tau(z)}$$

or

$$g_1(z) = \Psi_\tau(z) \cdot h_1(z),$$

where we denote $h_1(z) = g(M_\tau(z))$. Since, $h_1(0) = 0$ and $g_1 \in \mathrm{Star}(\Delta, \tau)$, we have that h_1 is also starlike, as proved above. But again $g(z) = h_1(M_\tau(z))$, and we are done. $\square$

Remark 5.5.1. Generally speaking we have proved the following: If g_1 and g_2 are two functions related by the formula:

$$g_1(M_\tau(z)) = g_2(z)\Psi_\tau(z), \tag{5.5.10}$$

then also:

$$g_2(M_\tau(z)) = g_1(z)\Psi_\tau(z). \tag{5.5.11}$$

In addition $g_1 \in \mathrm{Star}(\Delta, 0)$ if and only if $g_2 \in \mathrm{Star}(\Delta, \tau)$.

To prove Proposition 5.5.1 for the case when $\tau \in \partial\Delta$ we need a more detailed treatment of fanlike functions. We will do this separately in the next section.

5.6   A conjecture of Robertson and geometrical characteristics of fanlike functions

The first note related somehow to an application of the class $\mathrm{Fan}(\Delta)$ of starlike functions with respect to a boundary point to a boundary point seems to be due to E. Egervàry [35] in connection with Cesàro sums $\dfrac{1}{n}\sum_{k=1}^{n}(n - k + 1)z^k$ of the geometric series $z + z^2 + z^3 + \ldots + z^n + \ldots$. In particular, it turned out that the functions $h_n = 1 - \dfrac{2}{n(n+1)}\sum_{k=1}^{n}(n - k + 1)z^k$ belong to $\mathrm{Fan}(\Delta)$ with $h_n(1) = 0$.

Strangely, although the classes $\mathrm{Star}(\Delta, \tau)$ (of starlike functions with respect to interior points) have been studied by many mathematicians over a long period of time it seems that very few papers have been published up to 1981 on starlike functions with respect to a boundary point.

A breakthrough in this matter is due to M.S. Robertson [121], who suggested the inequality

$$\mathrm{Re}\left\{2\frac{zh'(z)}{h(z)} + \frac{1+z}{1-z}\right\} > 0, \quad z \in \Delta, \tag{5.6.1}$$

as a characterization criterion for those univalent holomorphic $h : \Delta \mapsto \mathbb{C}$ with $h(0) = 1$ such that $h(\Delta)$ is starlike with respect to the boundary point $h(1) := \lim_{r \to 1^-} h(r) = 0$ and lies in the right half-plane. This characterization was partially proved by Robertson himself under the additional assumption that h admits a holomorphic extension to a neighborhood of the closed unit disk. Furthermore, he established that this class is closely related to the class of close-to-convex functions. In particular, if h satisfying (5.6.1) is not a constant with $h(0) = 1$, then $g(z) = \log h(z)$, $\log h(0) = 0$, is close-to-convex with

$$\text{Re}\left[(1-z)^2 \frac{h'(z)}{h(z)}\right] < 0. \tag{5.6.2}$$

Note, in passing, that because of Proposition 5.2.2 inequality (5.6.2) is exactly a characterization of functions in the class $\text{Fan}(\Delta)$ normalized by the condition $h(1) := \lim_{r \to 1^-} h(r) = 0$.

Different applications of these results to convex functions were also exhibited in Robertson's work (see also [57]).

Observe, that the full proof of Robertson's conjecture was given by A. Lyzzaik [90], whilst a generalization of these results was later established by H. Silverman and E.M. Silvia [134].

In view of Carathéodory's Theorem of kernel convergence, a univalent function which is starlike with respect to a boundary point can be approximated by functions which are starlike with respect to interior points (see Figure 5.6 below). This approximation process can be considered dynamically as an evolution of the null points of these functions from the interior towards a boundary point. As mentioned above, this evolution is somehow connected to the evolution of semi-complete vector fields corresponding to starlike functions and the asymptotic behavior of one-parameter semigroups. So a natural question is how to trace these dynamics analytically in terms of inequalities (5.2.11′), (5.6.1) and (5.6.2).

In this chapter we will mostly follow the works [134] and [37]. We will show that condition (5.6.2) is equivalent to a generalized form of Robertson's condition (5.6.1). In addition, we will relate these conditions to some geometric considerations in the spirit of Silverman and Silvia [134].

We begin with the following observation. Let $g \in \text{Star}(\Delta, 0)$ and let $h \in \text{Hol}(\Delta, \mathbb{C})$ be given by

$$h(z) = \frac{(1-z)^2 g(z)}{z}. \tag{5.6.3}$$

Consider the functions

$$h_n(z) = H_{\tau_n}(z)\left(-g(z)\right),$$

where

$$H_{\tau_n}(z) = \frac{(z - \tau_n)(1 - z\overline{\tau_n})}{z}, \quad \tau_n \in \Delta,$$

are Hummel's multipliers. Since $-g$ also belongs to $\text{Star}(\Delta, 0)$ we have by Proposition 5.5.1 that $h_n \in \text{Star}(\Delta, \tau_n)$. Letting $\tau_n \in \Delta$ tend to 1 we obtain that h_n

uniformly approximates h on each compact subset of Δ, so $h(\Delta)$ is expected to be starshaped. In addition it follows by Hurwitz's theorem [68] that h is univalent.

Thus one may conjecture that condition (5.6.3) is also a characterization of those univalent holomorphic $h : \Delta \mapsto \mathbb{C}$ such that $h(\Delta)$ is starshaped with $h(1) := \lim_{r \to 1^-} h(r) = 0$, and hereby should be related to conditions (5.6.1) and (5.6.2).

Indeed, if (5.6.1) holds then setting (because of (5.6.3)) $g(z) = \dfrac{z}{(1-z)^2} h(z)$ we obtain

$$
\begin{aligned}
\operatorname{Re} \frac{zg'(z)}{g(z)} &= \operatorname{Re} \left\{ \frac{zh'(z)}{h(z)} + \frac{1+z}{1-z} \right\} \\
&= \frac{1}{2} \operatorname{Re} \left\{ 2\frac{zh'(z)}{h(z)} + \frac{1+z}{1-z} \right\} + \frac{1}{2} \operatorname{Re} \frac{1+z}{1-z} > 0,
\end{aligned}
$$

hence $g \in \operatorname{Star}(\Delta, 0)$. This sketches another proof that an $h \in \operatorname{Hol}(\Delta, \mathbb{C})$ satisfying (5.6.1) ought to belong to $\operatorname{Fan}(\Delta)$.

At the same time, if we keep in mind that (5.6.1) characterizes those $h \in \operatorname{Fan}(\Delta)$ with $h(0) = 1$, such that $h(\Delta)$ lies in the right half-plane, we reason that a stronger conjecture, namely that (5.6.1) is equivalent to (5.6.3), should be refuted.

So it might be worthwhile to replace (5.6.3) by a more qualified condition as well as to replace (5.6.1) by a generalized inequality, which are both related to the same geometrical location of the image $h(\Delta)$.

Following [134] for $\lambda \in [0,1)$ we define *the class G_λ of nonvanishing holomorphic functions $h : \Delta \mapsto \mathbb{C}$ with $h(0) = 1$ which satisfy the condition:*

$$
\operatorname{Re} \left[\frac{1}{1-\lambda} \frac{zh'(z)}{h(z)} + \frac{1+z}{1-z} \right] > 0 \quad \text{for all } z \in \Delta. \tag{5.6.4}
$$

First we will study some properties of the classes G_λ,

Lemma 5.6.1 *For each $\lambda \in [0,1)$ the set*

$$
\bigcup_{n=1}^{\infty} \left\{ \prod_{j=1}^{n} \left(\frac{1-z}{1-z\bar{\zeta}_j} \right)^{\lambda_j} : \sum_{j=1}^{n} \lambda_j = 2 - 2\lambda, \ |\zeta_j| = 1 \right\}
$$

is dense in G_λ in the topology of uniform convergence on compact subsets of Δ.

Proof. Since the function

$$
p(z) = \frac{1}{1-\lambda} \frac{zh'(z)}{h(z)} + \frac{1+z}{1-z}
$$

belongs to the class of Carathéodory: $\{ p \in \operatorname{Hol}(\Delta, \mathbb{C}), \ p(0) = 1, \ \operatorname{Re} p(z) > 0, \ z \in \Delta \}$, it follows from the Riesz–Herglotz representation of this class that:

$$
p(z) = \int_{|\zeta|=1} \frac{1 + z\bar{\zeta}}{1 - z\bar{\zeta}} \, dm(\zeta), \quad z \in \Delta,
$$

where m is a probability measure on the unit circle. Approximating the left hand side of the latter equation by the integral sums:

$$\sum_{j=1}^{n} \frac{1 + z\bar{\zeta}_j}{1 - z\bar{\zeta}_j} \Delta m_j, \quad \zeta_j \in \partial\Delta, \quad \sum_{j=1}^{n} \Delta m_j = 1,$$

and solving the initial value problem

$$\frac{1}{1 - \lambda} \frac{z h'_n(z)}{h_n(z)} = s(z), \quad h_n(0) = 1,$$

where

$$s(z) = \sum_{j=1}^{n} \left(\frac{1 + z\bar{\zeta}_j}{1 - z\bar{\zeta}_j} - \frac{1 + z}{1 - z} \right) \Delta m_j = 2 \sum_{j=1}^{n} \frac{\bar{\zeta}_j - 1}{(1 - z\bar{\zeta}_j)(1 - z)} \Delta m_j,$$

we obtain our assertion with $\lambda_j = 2(1 - \lambda)\Delta m_j$, $j = 1, \ldots, n$. $\square$

To continue we recall that a holomorphic function $g : \Delta \mapsto \mathbb{C}$ is said to be starlike of order $\lambda \in [0, 1)$ if $g(0) = 0$, $g'(0) = 1$ and $\mathrm{Re}\left(\frac{z g'(z)}{g(z)} \right) > \lambda$ for all $z \in \Delta$ (see Section 5.2). The set of such functions will be denoted by $S^*(\lambda)$.

The foregoing simple but important fact is due to Silverman and Silvia [134].

Lemma 5.6.2 Let $\lambda \in [0, 1)$ and let g and h be holomorphic functions in Δ related by the equation

$$g(z) := z(1 - z)^{2\lambda - 2} h(z). \tag{5.6.5}$$

Then $g \in S^*(\lambda)$ if and only if $h \in G_\lambda$.

Proof. This fact follows immediately by the equation

$$\frac{z g'(z)}{g(z)} - \lambda = (1 - \lambda) \left\{ \frac{1}{1 - \lambda} \frac{z h'(z)}{h(z)} + \frac{1 + z}{1 - z} \right\} \tag{5.6.6}$$

and the relations $g(0) = 0$, $g'(0) = h(0)$. $\square$
2mm

Lemma 5.6.3 If $0 \leq \lambda_1 \leq \lambda_2 < 1$, then $G_{\lambda_2} \subseteq G_{\lambda_1}$.

Proof. Assuming that $h \in G_{\lambda_2}$ we obtain by Lemma 5.6.2 that g defined by (5.6.5) with $\lambda = \lambda_2$, belongs to $S^*(\lambda_2)$. Let us now denote

$$\tilde{g}(z) = \frac{g(z)}{(1 - z)^{2(\lambda_2 - \lambda_1)}}. \tag{5.6.7}$$

We have by (5.6.5)

$$\tilde{g}(z) = g(z) := z(1 - z)^{2\lambda_2 - 2} h(z) = z(1 - z)^{2\lambda_1 - 2} h(z).$$

Thus, again in view of Lemma 5.6.2 we have to show that $\widetilde{g} \in S^*(\lambda_1)$. Indeed, on account of (5.6.7) we calculate

$$\operatorname{Re}\left(\frac{z\widetilde{g}'(z)}{\widetilde{g}(z)}\right) = \operatorname{Re}\left(\frac{zg'(z)}{g(z)}\right) - 2(\lambda_2 - \lambda_1)\operatorname{Re}\frac{z}{z-1}$$

$$\geq \lambda_2 - 2(\lambda_2 - \lambda_1)\operatorname{Re}\frac{z}{z-1} \geq \lambda_1,$$

$$\widetilde{g}(0) = 0, \qquad \widetilde{g}'(0) = 1,$$

and we have completed the proof. $\square$

Exercise 1. Show that $\operatorname{Re}\dfrac{z}{z-1} \leq \dfrac{1}{2}$, $z \in \Delta$.

Exercise 2. Prove that if $0 \leq \lambda_1$, $\lambda_2 < 1$, then $h \in G_{\lambda_1}$ if and only if

$$[h(z)]^{\frac{1-\lambda_2}{1-\lambda_1}} \in G_{\lambda_2}.$$

A natural question related to the last lemma is: a given h of the class G_λ for some $\lambda \in [0,1)$ find the maximal $\lambda = \lambda^*$, such that h also belongs to G_{λ^*}. In other words, G_{λ^*} should be the minimal class which contains h. It turns out, that this question is closely connected to another one: if h is a starlike function with respect to a boundary point, how does one determine the minimal angle θ such that $h(\Delta)$ lies in the wedge of this angle.

In [37] it was discussed how to resolve the above questions.

Proposition 5.6.1 *Let* $h : \Delta \mapsto \mathbb{C}$ *be holomorphic and let* $\lambda \in [0,1)$. *If* h *is not a constant and* $h(0) = 1$, *then the following conditions are equivalent.*

(i) $\operatorname{Re}\left[\dfrac{1}{1-\lambda}\dfrac{zh'(z)}{h(z)} + \dfrac{1+z}{1-z}\right] > 0$ *for all* $z \in \Delta$.

(ii) There exists a starlike function $g : \Delta \mapsto \mathbb{C}$ *of order* λ *such that*

$$h(z) = \frac{(1-z)^{2-2\lambda}\,g(z)}{z}, \qquad z \in \Delta.$$

(iii) The function h *belongs to* $\operatorname{Fan}(\Delta)$ *with* $h(1) := \lim\limits_{r \to 1^-} h(r) = 0$ *and* $h(\Delta)$ *lies in a wedge of the angle* $2\pi(1 - \lambda)$.

(iv) The function h *is a univalent function on* Δ *such that* $f(z) := h(z)/h'(z)$, $z \in \Delta$, *is a semi-complete vector field with*

$$\angle \lim_{z \to 1} \operatorname{Re}\frac{f(z)}{z-1} \geq \frac{1}{2-2\lambda},$$

where the limit in the left hand side of the latter inequality exists finitely.

Moreover, the equality in (iv) can be reached if and only if $\lambda = \lambda^*$, *where* G_{λ^*} *is the minimal class of* G_λ *which includes* h, *and if and only if the wedge of the angle* $2\pi(1 - \lambda^*)$ *is the smallest one which contains* $h(\Delta)$.

Proof. The equivalence of conditions (i) and (ii) is the content of Lemma 5.6.2. Our next steps are slightly simpler than in [37].

First we claim that condition (ii) and Lemma 5.6.3 imply that $h \in G_\lambda \setminus \{1\}$ must be univalent. Indeed, setting in this lemma $\lambda_1 = 0$, $\lambda_2 = \lambda \geq 0$ we obtain that h admits a similar representation as (5.6.3):

$$h(z) = \frac{(1-z)^2 \widetilde{g}(z)}{z} \tag{5.6.3'}$$

with some $\widetilde{g} \in S^*$. Then, as mentioned above, the approximation process:

$$h_n(z) = \frac{(\tau_n - z)(1 - z\overline{\tau_n})}{z} \widetilde{g}(z), \quad \tau_n \in \Delta, \ \tau_n \to 1,$$

implies the desired claim.

Next we want to show that h belongs to Fan(Δ). In view of Proposition 5.2.2 this fact will be proved once we show that

$$\mathrm{Re}\left((1-z)^2 \frac{h'(z)}{h(z)}\right) \leq 0, \quad z \in \Delta. \tag{5.6.8}$$

To this end let $0 < r < 1$, and define $h_r : \Delta \mapsto \mathbb{C}$ by

$$h_r(z) := h(rz)\left(\frac{1-z}{1-rz}\right)^{2(1-\lambda)}, \quad z \in \Delta.$$

If we use the corresponding function $g \in S^*(\lambda)$ we can write equivalently that

$$h_r(z) = \left(\frac{g(rz)}{r}\right)\left(\frac{1}{z}\right)(1-z)^{2(1-\lambda)}.$$

This last representation of h_r shows that it belongs to G_λ. Its definition shows that $h_r \to h$ as $r \to 1^-$ and that h_r is continuous on the closed disk $\overline{\Delta}$. Therefore the claimed inequality will follow if we inspect it for h_r and for $z = e^{i\varphi} \in \partial\Delta$.

Indeed, for such z we have

$$\mathrm{Re}\left((1-z)^2 \frac{h_r'(z)}{h_r(z)}\right)$$

$$= \mathrm{Re}\left[\frac{(1-z)^2}{z}\left(\frac{zh_r'(z)}{h_r(z)} + (1-\lambda)\frac{1+z}{1-z}\right) + \frac{(1-z)^2}{z}(1-\lambda)\frac{1+z}{1-z}\right]$$

$$= \mathrm{Re}\left[(z - 2 + \bar{z})\left(\frac{zh_r'(z)}{h_r(z)} + (1-\lambda)\frac{1+z}{1-z}\right) + (1-\lambda)(\bar{z} - z)\right]$$

$$= 2(\cos\varphi - 1)\mathrm{Re}\left[\frac{zh_r'(z)}{h_r(z)} + (1-\lambda)\frac{1+z}{1-z}\right] \leq 0,$$

as claimed.

The fact established in (5.6.8) and the Berkson–Porta formula mean, actually, that $f : \Delta \mapsto \mathbb{C}$, defined by

$$f(z) = \frac{h(z)}{h'(z)}$$

is a semi-complete vector field. In addition, the same formula implies (because of the uniqueness property) that $\tau = 1$ must be the sink point of the semigroup $S = \left\{ F_t = h^{-1}\left(e^{-t}h(z)\right)\right\}_{t\geq 0}$ generated by f (see Section 5.2).

To proceed with the proof of the implication (ii)$\Rightarrow$(iii) we show now that $\lim\limits_{r\to 1^-} h(r) = 0$. In fact, we intend to prove a more general condition:

$$\angle \lim_{z\to 1} h(z) = 0. \tag{5.6.9}$$

Fix any element z_0 in Δ and define $z_t = F_t(z_0)$, $t \geq 0$. We have $z_t \to 1$ and

$$h(z_t) = e^{-t}h(z) \to 0, \text{ as } t \to \infty. \tag{5.6.10}$$

Thus what we need to show is that h is bounded in any nontangential approach region

$$\Gamma(1,k) = \{z \in \Delta : |1 - z| \leq k(1 - |z|), \ k > 1\}\,.$$

Returning to the representation (5.6.3') the latter fact is easily seen with the help of Koebe distortion theorem:

$$|h(z)| = \left|\frac{(1-z)^2}{z}\right| |\widetilde{g}(z)| \leq \frac{|1-z|^2}{|z|} \frac{|z|}{(1-|z|)^2} \leq k^2,$$

whenever $z \in \Gamma(1,k)$. Now by the Lindelöf's Principle with (5.6.10) valid (5.6.9) results.

To accomplish (iii) we need to show that $h(\Delta)$ lies in a wedge of the angle $2\pi(1 - \lambda)$.

By Lemma 5.6.1 we may assume that h is of the form

$$h(z) = \prod_{j=1}^{n}\left(\frac{1-z}{1-z\bar{\zeta_j}}\right)^{\lambda_j},$$

where $|\zeta_j| = 1$, $\zeta_j \neq 1$ and $\sum\limits_{j=1}^{n}\lambda_j = 2(1 - \lambda)$. Each function $\omega_j(z) := \dfrac{1-z}{1-z\bar{\zeta_j}}$ maps the open unit disk Δ onto a half-plane. In other words, $\text{Re}\left(e^{i\beta_j}\omega_j(z)\right) > 0$ for some β_j.

Denoting $\sum\limits_{j=1}^{n}\lambda_j\beta_j$ by β, we have for each $z \in \Delta$

$$\left|\arg e^{i\beta}h(z)\right| = \left|\arg e^{i\beta}\prod_{j=1}^{n}\omega_j^{\lambda_j}\right| = \left|\sum_{j=1}^{n}\lambda_j\left(\arg e^{i\beta_j}\omega_j\right)\right|$$

$$< \sum_{j=1}^{n}\lambda_j\left(\frac{\pi}{2}\right) = \pi(1 - \lambda).$$

Hence $g(\Delta)$ is contained in a wedge of the angle $2\pi(1 - \lambda)$ as claimed.

Now following the idea suggested in [90] we will show that (iii)$\Rightarrow$(ii). Let $h_0(z) = h(z)^{\frac{1}{1-x}}$. Then $h_0(0) = 1$, $h_0(1) = 0$, h_0 is univalent and $h_0(\Delta)$ is starshaped with respect to $h_0(1) = 0$. Set

$$D_n = h_0(\Delta) \cup \left\{ z \in \mathbb{C} : |z| < \frac{1}{n} \right\}, \quad n = 1, 2, \ldots,$$

and for each n let $h_n : \Delta \mapsto D_n$ be the conformal mapping of Δ onto D_n such that $h_n(0) = 1$ and $\arg h'_n(0) = \arg h'_0(0)$ (see Figure 5.6).

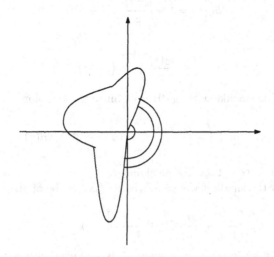

Figure 5.6: An approximation of a starshaped domain with respect to a boundary point.

By Carathéodory's Kernel Theorem we know that

$$\lim_{n \to \infty} h_n = h_0,$$

uniformly on each compact subset of Δ. Since each $h_n(\Delta)$ is starshaped there are starlike functions g_n with $g_n(0) = 0$ and numbers τ_n, $|\tau_n| < 1$, such that

$$h_n(z) = \frac{g_n(z)}{z}(z - \tau_n)(1 - \overline{\tau_n}z), \quad z \in \Delta,$$

(see Section 5.5).

Note that $1 = h_n(0) = -\tau_n g'_n(0)$ and that

$$h'_n(0) = \frac{1}{2} \frac{g''_n(0)}{g'_n(0)} + g'_n(0)(1 + |\tau_n|^2)$$

for all n. If the sequence $\{g'_n(0)\}$ had been unbounded then we would have reached a contradiction, because $h'_n(0) \to h'_0(0)$ and $\left| g''_n(0)/g'_n(0) \right| \leq 4$.

Thus $\{g_n'(0)\}$ is bounded and we can extract a convergent subsequence of $\{g_n\}$. It is clear that we can assume that the corresponding subsequence of $\{\tau_n\}$ converges to a point $\tau \in \overline{\Delta}$. Denoting the limit function of the convergent subsequence of $\{g_n\}$ by g_0, we see that

$$h_0(z) = \frac{g_0(z)}{z}(z - \tau)(1 - \bar{\tau}z), \quad z \in \Delta.$$

Letting z approach 1 we conclude that $\tau = 1$. Hence

$$h_0(z) = \left(-\frac{g_0(z)}{z}\right)(1 - z)^2$$

and

$$h(z) = \left(-\frac{g_0(z)}{z}\right)^{1-\lambda}(1 - z)^{2-2\lambda},$$

where $g_0 : \Delta \mapsto \mathbb{C}$ is starlike with $g_0(0) = 0$. Since the function

$$g(z) := z\left(-\frac{g_0(z)}{z}\right)^{1-\lambda} = z(1 - z)^{2\lambda-2}h(z)$$

is starlike of order λ we obtain (ii), as claimed.

(iii)$\Rightarrow$(iv). Let the smallest wedge in which $h(\Delta)$ lies be of an angle $2\pi(1-\lambda^*)$. Then $\lambda^* \geq \lambda$,

$$\mathrm{Re}\left[\frac{1}{1-\lambda^*}\frac{zh'(z)}{h(z)} + \frac{1+z}{1-z}\right] > 0, \quad z \in \Delta,$$

and this inequality no longer holds when λ^* is replaced with any number $\lambda^* < \lambda < 1$. By the Riesz–Herglotz representation theorem we can write

$$\frac{1}{1-\lambda_1}\frac{zh'(z)}{h(z)} + \frac{1+z}{1-z} = \int_{|\zeta|=1}\frac{1+z\bar{\zeta}}{1-z\bar{\zeta}}dm(\zeta),$$

where m is a probability measure on the unit circle. After some calculations we obtain

$$h(z) = (1-z)^{2(1-\lambda^*)}\exp\left(-2(1-\lambda^*)\int_{|\zeta|=1}\log(1 - z\bar{\zeta})dm(\zeta)\right).$$

Again we note that (5.6.9) no longer holds when λ^* is replaced with any number $\lambda^* < \lambda < 1$. Let δ denote the Dirac measure at $\zeta = 1 \in \partial\Delta$. Decomposing m relative to δ, we can write $m = (1 - a)\nu + a\delta$, where $0 \leq a \leq 1$, and ν and δ are mutually singular probability measures. It follows that

$$h(z) = (1-z)^{2(1-\lambda)}\exp\left(-2(1-\lambda)\int_{|\zeta|=1}\log(1 - z\bar{\zeta})d\nu(\zeta)\right),$$

where $\lambda = 1 - (1 - \lambda^*)(1 - a)$.

If $a > 0$ we reach a contradiction because $\lambda > \lambda^*$. Thus $a = 0$ and $m = \nu$.

Let $g = h/h'$. Then f is semi-complete by Proposition 5.2.4. Using (5.6.8) or (5.6.9) we see that

$$\frac{z-1}{f(z)} = 2(1 - \lambda^*) \int_{|\varsigma|=1} \frac{1 - \bar{\varsigma}}{1 - z\bar{\varsigma}} \, d\nu(\varsigma), \quad z \in \Delta.$$

Let $\{z_n\}$ be any sequence in

$$\Gamma(1, k) = \{z \in \Delta : |1 - z| \le k\,(1 - |z|)\,, \; k > 1\}.$$

which tends to 1. Consider the functions $h_n : \partial\Delta \mapsto \mathbb{C}$, $n = 1, 2, \ldots$, defined by

$$h_n(\varsigma) := \frac{1 - \bar{\varsigma}}{1 - z_n\varsigma}, \quad \varsigma \in \partial\Delta.$$

Since the function h_n maps the unit circle $\partial\Delta$ onto the circle $|\xi - c_n| = |c_n|$, where $c_n = (1 - \bar{z}_n)/(1 - |z_n|^2)$, $n = 1, 2, \ldots$, we obtain that

$$|h_n(\varsigma)| \le 2|c_n| \le 2k.$$

Using (5.6.10) and applying Lebesgue's Bounded Convergence Theorem we now obtain

$$\angle \lim_{z \to 1} \frac{z-1}{f(z)} = 2(1 - \lambda^*) \lim_{n \to \infty} \int_{|\varsigma|=1} h_n(\varsigma) d\nu(\varsigma)$$

$$= 2(1 - \lambda^*) \le 2(1 - \lambda). \tag{5.6.11}$$

In other words, condition (d) holds.

Finally, we show the implication (iv)$\Rightarrow$(iii). Note that by Proposition 4.6.2, $\tau = 1$ is the sink point of the semigroup generated by f and $\angle \lim_{z \to 1} f'(z)$ is, in fact, a real number. Therefore

$$\angle \lim_{z \to 1} \frac{f(z)}{z-1} \ge \frac{1}{2(1 - \lambda)}.$$

Moreover, $f(z) = \dfrac{h(z)}{h'(z)} = -(1 - z)^2 p(z)$, where $p : \Delta \mapsto \mathbb{C}$ is holomorphic with $\operatorname{Re} p(z) \ge 0$ for all $z \in \Delta$. Again applying Proposition 5.2.4 and repeating the arguments as in the proof of (5.6.9) we obtain that h is starlike with respect to a boundary point with $\lim_{r \to 1} h(r) = 0$. Let the smallest wedge in which $h(\Delta)$ lies be of an angle $2\pi(1 - \lambda^*)$, where $\lambda^* \in [0, 1)$. As we saw in the proof of implication (iii)$\Rightarrow$(iv), it follows that

$$\angle \lim_{z \to 1} \frac{f(z)}{z-1} = \frac{1}{2(1 - \lambda^*)}. \tag{5.6.12}$$

Comparing the latter equality with (5.6.11) we see that $\lambda \leq \lambda^*$. Thus $h(\Delta)$ lies in a wedge of an angle $2\pi(1 - \lambda)$, as claimed. This concludes the proof of our assertion. $\square$

Remark 5.6.1. Thus given $h \in \mathrm{Fan}(\Delta)$ with $\lim\limits_{r \to 1} h(r) = 0$, formula (5.6.12) infers that the value of smallest angle θ such that $h(\Delta)$ lies in its wedge is π multiplied by the angular limit of the Visser–Ostrowski quotient:

$$\theta = \pi \angle \lim_{z \to 1} \frac{(z - 1)h'(z)}{h(z)}.$$

Corollary 5.6.1 *If $h \in \mathrm{Fan}(\Delta)$ with $\lim\limits_{r \to 1} h(r) = 0$ satisfies the Visser–Ostrowski condition:*

$$\angle \lim_{z \to 1} \frac{(z - 1)h'(z)}{h(z)} = 1$$

(in particular, if h is conformal, or, more generally, isogonal at 1), then the smallest wedge which contains $h(\Delta)$ is precisely the right half-plane

$$\Pi_+ = \{z \in \mathbb{C} : \mathrm{Re}\, z \geq 0\}.$$

Remark 5.6.2 Since $G_\lambda \subset G_0$, it follows by the above proposition that $h \in \mathrm{Fan}(\Delta)$ with $\lim\limits_{r \to 1^-} h(r) = 0$ if and only if it satisfies the equation

$$h(z) = \frac{-(1 - z)^2}{z} g(z),$$

where $g \in \mathrm{Star}(\Delta, 0)$. This proves Proposition 5.5.1 (Hummel's representation formula) for the case $\tau = 1$. Geometrically this fact should be understood as $h(\Delta)$ lies in the wedge of angle 2π.

5.7 Converse theorems on starlike, spirallike and fanlike functions

In this section we consider *inter alia* the following converse problem: given $f \in \mathcal{G}\,\mathrm{Hol}(\Delta)$ find conditions such that the equation

$$h(z) = h'(z)f(z) \tag{5.7.1}$$

has a global solution in Δ which is univalent (consequently, starlike) in Δ.

If $f(\tau) = 0$ for some $\tau \in \Delta$ and we are looking for the solution of (5.7.1) satisfying the initial conditions:

$$h(\tau) = 0 \text{ and } h'(\tau) \neq 0, \qquad (5.7.2)$$

then the necessary restriction is that $f'(\tau) = 1$.

More generally, if $f'(\tau) = \mu \neq 0$ then instead of equation (5.7.1) we must consider equation:

$$\mu h(z) = h'(z) \cdot f(z) \qquad (5.7.3)$$

with initial conditions (5.7.2), which determines a spirallike function h.

Proposition 5.7.1 *Let* $f \in \mathcal{G} \operatorname{Hol}(\Delta)$ *be such that:*

$$f(0) = 0 \qquad (5.7.4)$$

and

$$f'(0) = \mu. \qquad (5.7.5)$$

Suppose that $h \in \operatorname{Hol}(\Delta)$ *is a solution of (5.7.3). Then:*

$$h(z) = h'(0) \lim_{t \to \infty} e^{\mu t} \cdot F_t(z), \qquad (5.7.6)$$

where $\{F_t\}_{t>0}$ *is the semigroup, generated by* f.

Proof. Let $\{F_t\}_{t \geq 0} = S_f$ be a semigroup generated by f. Then it follows by (5.7.3) that

$$\mu h(F_t(z)) = h'(F_t(z))f(F_t(z)) = -h'(F_t(z)) \cdot \frac{\partial F_t(z)}{\partial t}$$

or

$$\mu G_t(z) = -\frac{\partial G_t(z)}{\partial t}, \qquad (5.7.7)$$

where $G_t(z) = h(F_t(z))$, $t \geq 0$, $z \in \Delta$ and

$$G_0(z) = h(z). \qquad (5.7.8)$$

Solving (5.7.7) with initial data (5.7.8) we obtain:

$$G_t(z) = e^{-\mu t} \cdot h(z)$$

or

$$h(F_t(z)) = e^{-\mu t} \cdot h(z). \qquad (5.7.9)$$

Consequently, to prove our assertion, it is sufficient to show that

$$\lim_{t \to \infty} e^{\mu t} [h(F_t(z)) - h'(0)F_t(z)] = 0 \qquad (5.7.10)$$

for all $z \in \Delta$.

Note, that by Taylor's formula we have that for each $w \in \Delta$

$$h(w) - h'(0)w = \sum_{n=2}^{\infty} \frac{1}{n!} \frac{d^n h}{dz^n}(0) \, w^n = \tilde{h}(w).$$

If $M = \sup_{|w|<1} \tilde{h}(w)$ then the Schwarz Lemma (see Section 1.1) implies

$$\left| \tilde{h}(w) \right| \leq M \, |w|^2$$

for all $w \in \Delta$.

Setting $w = F_t(z)$ we obtain

$$\left| \tilde{h}(F_t(z)) \right| \leq M \, |F_t(z)|^2. \tag{5.7.11}$$

In addition, we know that

$$\frac{|F_t(z)|}{(1 - |F_t(z)|)^2} \leq \exp\left(- \operatorname{Re} \mu t\right) \frac{|z|}{(1 - |z|)^2}$$

(see Proposition 4.1.1).

Hence (5.7.11) implies

$$\left| \tilde{h}(F_t(z)) \right| \leq \frac{M \, |z|^2}{(1 - |z|)^4} \cdot \exp(-2 \operatorname{Re} \mu t),$$

since $(1 - |F_t(z)|)^2 \leq 1$.

Finally, since $\operatorname{Re} \mu > 0$ we have

$$\left| e^{\mu t} \tilde{h}(F_t(z)) \right| = \exp(\operatorname{Re} \mu t) \left| \tilde{h}(F_t(z)) \right|$$

$$\leq \frac{M \, |z|^2}{(1 - |z|)^4} \cdot \exp(- \operatorname{Re} \mu t) \to 0$$

as $t \to \infty$ and the result follows. $\square$

Thus if equation (5.7.3) has a solution $h \in \operatorname{Hol}(\Delta)$ satisfying the condition

$$h'(0) = k \neq 0, \tag{5.7.12}$$

then there is no other solution satisfying the same condition.

Conversely, if the limit

$$h(z) := k \lim_{t \to \infty} e^{\mu t} F_t(z)$$

exists it is easy to see that $h(z)$ satisfies (5.7.3) with (5.7.12). Indeed, in this case

$$h'(z)f(z) = k \lim_{t \to \infty} e^{t\mu} \cdot \frac{\partial F_t(z)}{\partial z} \cdot f(z) = k \lim_{t \to \infty} e^{t\mu} f(F_t(z))$$

$$= k \lim_{t \to \infty} e^{t\mu} \left[\mu F_t(z) + \sum_{n=2}^{\infty} \frac{1}{n!} \frac{d^n f}{dz^n}(0)(F_t(z)^n) \right].$$

Again, as above, it can be shown that

$$\lim_{t\to\infty}\sum_{n=2}^{\infty}\frac{1}{n!}\frac{d^n f}{dz^n}(0)(F_t(z)^n)=0.$$

Hence, we obtain:

$$h'(z)f(z)=\mu k\lim_{t\to\infty}e^{t\mu}F_t(z)=\mu h(z)$$

and we have completed the proof. $\square$

Thus to solve the problem (5.7.3)–(5.7.12) we only need to show that the limit in (5.7.6) exists.

To this end we define

$$u(t,z)=e^{t\mu}F_t(z)\quad\text{for all }t\geq0\text{ and }z\in\Delta.$$

Then for fixed $z\in\Delta$,

$$\begin{aligned}\frac{\partial u(t,z)}{\partial t}&=\mu e^{\mu t}F_t(z)-e^{\mu t}f(F_t(z))\\&=\mu u(t,z)-e^{\mu t}f(e^{-\mu t}u(t,z))\\&=e^{\mu t}\left(\mu e^{-\mu t}u(t,z)-f(e^{-\mu t}u(t,z))\right)\end{aligned}$$

and $u(0,z)=z$.

Therefore the function $u(t,z):\mathbb{R}^+\to\mathbb{C}$ satisfies the equation:

$$\begin{cases}\dfrac{\partial u(t,z)}{\partial t}+e^{t\mu}g(e^{-\mu t}u(t,z))=0,\\[2mm]u(0,z)=z,\end{cases}\tag{5.7.13}$$

where

$$g(z)=f(z)-\mu z=\sum_{n=2}^{\infty}\frac{1}{n!}\frac{d^n f}{dz^n}(0)\,z^n.\tag{5.7.14}$$

In turn, (5.7.13) implies:

$$u(t_1,z)-u(t_2,z)=-\int_{t_1}^{t_2}e^{t\mu}g(e^{-\mu t}u(t,z))\,dt$$

for each pair $t_1,t_2\in\mathbb{R}^+$ and $z\in\Delta$.

It follows by (5.7.14) that $g'(0)=0$. Hence, for each $z\in\Delta$:

$$|g(z)|\leq M_1\,|z|^2,$$

where $M_1=\sup_{z\in\Delta}|g(z)|$. Therefore

$$\begin{aligned}|g(e^{-\mu t}u(t,z))|&=|g(F_t(z))|\leq M_1|F_t(z)|^2\\&\leq M_1e^{-2\,\mathrm{Re}\,\mu t}\frac{|z|^2}{(1-|z|)^4}.\end{aligned}$$

Using (5.7.6) we obtain for each $z \in \Delta$:

$$|u(t_1, z) - u(t_2, z)| \leq M_1 \int_{t_1}^{t_2} e^{-\operatorname{Re}\mu t} dt \cdot \frac{|z|^2}{1 - |z|^4} \to 0$$

as t_1 and t_2 tend to ∞.

Therefore the limit

$$\lim_{t \to \infty} u(t, z) = \lim_{t \to \infty} e^{t\mu} F_t(z)$$

exists. $\square$

Thus we have proved the following assertion.

Proposition 5.7.2 *Let $f \in \mathcal{G} \operatorname{Hol}(\Delta)$ be a semi-complete vector field which satisfies the conditions: $f(0) = 0$, and $f'(0) = \mu$. Then the equation*

$$\mu h(z) = h'(z) \cdot f(z)$$

has a unique solution $h \in \operatorname{Hol}(\Delta)$ which satisfies the conditions: $h(0) = 0$, $h'(0) = k \neq 0$. This solution is a univalent spirallike function and it has the form:

$$h(z) = k \lim_{t \to \infty} e^{t\mu} F_t(z),$$

where $\{F_t\}$ is the semigroup, generated by f.

In particular, when μ is a real number $h(z)$ is a starlike function.

Turning to a more general situation, assume now that

$$f(z) = (z - \tau)(1 - z\bar{\tau})q(z), \qquad (5.7.15)$$

with $\operatorname{Re} q(z) > 0$ everywhere. We already know that for each $f \in \mathcal{G} \operatorname{Hol}(\Delta)$ with an interior null point τ representation (5.7.15) holds.

In this case:

$$f'(\tau) = (1 - |\tau|^2)q(\tau). \qquad (5.7.16)$$

If we define

$$g(z) = (M_\tau)'(M_\tau(z)) f(M_\tau(z)),$$

we obtain $g(0) = 0$ and $\mu = g'(0) = (1 - |\tau|^2)q(\tau) = f'(\tau)$ and $g \in \mathcal{G} \operatorname{Hol}(\Delta)$.

Therefore the equation

$$\mu h_1(z) = h'_1(z) \cdot g(z)$$

with the initial conditions $h(0) = 0$, $h'_1(0) \neq 0$ has a unique solution:

$$h_1(z) = h'_1(0) \cdot \lim_{t \to \infty} e^{\mu t} G_t(z),$$

where $\{G_t(z)\}_{t>0} = S_g$ is the semigroup generated by g.

Define now

$$h(z) = h_1(M_\tau(z)), \qquad (5.7.17)$$

from which we obtain $h(\tau) = h_1(0) = 0$ and $h'(z) = h'_1(M_\tau(z)) \cdot (M_\tau)'(z)$.

In particular,

$$h'(\tau) = h_1'(0)(|t|^2 - 1)^{-1}. \tag{5.7.18}$$

In addition,

$$\begin{aligned} \mu h(z) &= \mu h_1(M_\tau(z)) = \mu \cdot h_1'(M_\tau(z))g(M_\tau(z)) \\ &= \mu h_1(M_\tau(z)) \cdot (M_\tau)'(z)f(z) = \mu h'(z)f(z). \end{aligned} \tag{5.7.19}$$

Note now that if $\{F_t\}_{t>0} = S_f$ is the semigroup generated by f, then $G_t(z) = M_\tau(F_t(M_\tau(z)))$ is the semigroup generated by g (recall, that M_τ is an involution), and we have two symmetric relations:

(i) $M_\tau(G_t(z)) = F_t(M_\tau(z))$;
(ii) $M_\tau(F_t(z)) = G_t(M_\tau(z))$.

Since, $h(z) = h_1(M_\tau(z))$ and $h_1(z) = h_1'(0) \cdot \lim\limits_{t\to\infty} e^{\mu t}G_t(z)$, we obtain:

$$\begin{aligned} h(z) &= h_1'(0) \cdot \lim_{t\to\infty} e^{\mu t}G_t(M_\tau(z)) \\ &= h'(\tau)(|\tau|^2 - 1) \cdot \lim_{t\to\infty} e^{\mu t}M_\tau(F_t(z)) \\ &= h'(\tau)(|\tau|^2 - 1) \cdot \lim_{t\to\infty} e^{\mu t}\frac{\tau - F_t(z)}{1 - \bar\tau F_t(z)} \\ &= h'(\tau) \lim_{t\to\infty} e^{\mu t}(F_t(z) - \tau). \end{aligned}$$

This formula gives us the solution of (5.7.3) under the conditions $h(\tau) = 0$, $h'(\tau) \neq 0$. $\square$

So, we arrive at the following conclusion:

Proposition 5.7.3 *Let* $f \in \mathcal{G}\,\mathrm{Hol}(\Delta)$ *have the form* (5.7.15)

$$f(z) = (z - \tau)(1 - z\bar\tau)q(z)$$

with $\mathrm{Re}\,q(z) > 0$, $\tau \in \Delta$ *and* $q(0) \neq 0$. *Then the equation*

$$\mu h(z) = h'(z) \cdot f(z)$$

with $\mu = (1 - |\tau|^2)q(0)$ *has a unique solution* $h(z)$ *satisfying the conditions:*

$$h(\tau) = 0, \qquad h'(\tau) = k \neq 0.$$

This solution is a univalent spirallike function on Δ *which can be defined by the formula:*

$$h(z) = k \lim_{t\to\infty} e^{\mu t}(F_t(z) - \tau).$$

If, in particular, $q(\tau)$ *is real then* h *is a starlike function on* Δ.

Now we consider the case when $f \in \mathcal{G}\,\mathrm{Hol}(\Delta)$ has no null point in Δ. It looks like this case is simpler, because equation (5.7.3) has no singularity in Δ, and one can define a local solution of (5.7.3) for each $\mu \in \mathbb{C}$ under a corresponding initial condition, say

$$h(0) = 1, \qquad (5.7.20)$$

and the boundary condition

$$\lim_{r \to 1^-} h(r\tau) = 0, \qquad (5.7.21)$$

where $\tau \in \partial\Delta$ is the sink point of the semigroup generated by f (see Sections 5.2 and 5.3). But, on the other hand, it is not clear why such a solution has an extension to all of Δ to be a univalent (spirallike) function (with respect to a boundary point). Moreover, if we wish this solution to satisfy the Visser–Ostrowski condition at the point $\tau \in \partial\Delta$ we must require for the number μ in (5.7.3) to be real.

Proposition 5.7.4 *Let f be a semi-complete vector field in Δ with no null point inside, and let $\tau \in \partial\Delta$ be the sink point of the semigroup generated by f. If $\beta = \angle \lim\limits_{z \to \tau} f'(z)$ then for a real $\mu > 0$ the problem*

$$\mu h(z) = h'(z)f(z), \ h(0) = 1 \qquad (5.7.22)$$

has a unique univalent solution in Δ if and only if

$$\mu \le 2\beta. \qquad (5.7.23)$$

   This solution is a fanlike function (i.e., a starlike function with respect to a boundary point), the image of which lies in the wedge of angle $\theta = \pi\mu/\beta$.

   Proof. Without loss of generality we can assume that $\tau = 1$. If $h \in \mathrm{Univ}(\Delta)$ satisfies (5.7.22) with some real $\mu > 0$, then its image $h(\Delta)$ must be starshaped by Proposition 5.2.4. Furthermore, if we define $f_1(z) = \dfrac{1}{\mu}f(z)$ we have

$$\beta_1 := \angle \lim_{z \to \tau} f_1'(z) = \frac{\beta}{\mu} > 0$$

and h satisfies the equation

$$h(z) = h'(z)f_1(z)$$

with $h(0) = 1$. Then it follows by Proposition 5.6.1 that there is $\lambda \in [0,1)$ such that h belongs to the class G_λ and $\beta_1 := \angle \lim\limits_{z \to \tau} f_1'(z) \ge \dfrac{1}{2 - 2\lambda} \ge \dfrac{1}{2}$. This implies (5.7.23). In addition the angular limit of the Visser–Ostrowski quotient is

$$\nu = \angle \lim_{z \to 1^-} \frac{h'(z)(z - 1)}{h(z)} = \frac{1}{\beta_1} = \frac{\mu}{\beta},$$

and the smallest wedge which contains $h(\Delta)$ is of the angle $\frac{\pi\mu}{\beta}$.

   Conversely. To solve (5.7.22) we first consider the problem:

$$\beta \widetilde{h}(z) = \widetilde{h}'(z)f(z), \ \widetilde{h}(0) = 1, \qquad (5.7.24)$$

where $\beta = \angle \lim\limits_{z \to 1^-} f'(z)$.

Consider the functions:

$$f_n(z) = \frac{1}{n}z + f(z), \quad n = 1, 2, \ldots . \tag{5.7.25}$$

Firstly, it is known that $f_n \in \mathcal{G}\,\mathrm{Hol}(\Delta)$ for all $n = 1, 2, \ldots$, since the class $\mathcal{G}\,\mathrm{Hol}(\Delta)$ is a real cone. Secondly, for each $n = 1, 2, \ldots$ the equation

$$f_n(z) = 0 \tag{5.7.26}$$

has a unique solution $\tau_n \in \Delta$ such that $\tau_n \to 1^-$, as $n \to \infty$.

Indeed, equation (5.7.26) is equivalent to the following one

$$z + nf(z) = 0,$$

which defines the values of the resolvent J_n at the point zero, i.e., $\tau_n = J_n(0)$.

If we denote $\mu_n = \dfrac{1}{n} + f'(\tau_n)$ we obtain that $\mu_n \to \beta$ as $n \to \infty$ and $f'_n(\tau_n) = \mu_n$. Therefore, by the above Proposition 5.7.3, for each $n = 1, 2, \ldots$ the equation

$$\mu_n h_n(z) = h'_n(z) \cdot f_n(z) \tag{5.7.27}$$

has a univalent solution $h_n(z)$ determined by $h'_n(\tau_n) \neq 0$.

Since $\tau_n \neq 0$ for all $n = 1, 2, \ldots$ and h_n is univalent we have $h_n(0) \neq 0$ for all $n = 1, 2, \ldots$. Therefore, we can define the functions:

$$\tilde{h}_n(z) = \frac{1}{h_n(0)} \cdot h_n(z)$$

which also satisfy equation (5.7.27), with $\tilde{h}'_n(\tau_n) = \frac{1}{h_n(0)} \cdot h'_n(\tau_n)$.

In addition,

$$\tilde{h}_n(0) = 1 \quad n = 1, 2, \ldots . \tag{5.7.28}$$

Now for each $r \in [0, 1)$ we can find $N > 0$ such that for all $n > N$, $|\tau_n| > r$, that is f_n do not vanish on the disk $|z| \leq r$. Therefore for such z (i.e., $|z| \leq r$) we can write by using (5.7.27) and (5.7.28):

$$\tilde{h}_n(z) = \exp\left\{ \mu_n \cdot \int_0^z \frac{dz}{f_n(z)} \right\}.$$

Furthermore, since $f(0) \neq 0$ there is a neighborhood U of the point $z = 0$ in which (5.7.24) has a unique solution:

$$\tilde{h}(z) = \exp\left\{ \mu \cdot \int_0^z \frac{dz}{f(z)} \right\}. \tag{5.7.29}$$

Since $\mu_n \to \beta$ and $f_n(z) \to f(z)$ for all $z \in \Delta$, we have that $\tilde{h}_n(z)$ converges to $h(z)$ in this neighborhood. It then follows by the Vitali theorem that $\tilde{h}_n(z)$

converges to $\widetilde{h}(z)$ on all of the disk $|z| < r$. Since r is arbitrary we obtain that $\widetilde{h}(z)$ is well defined on all of Δ. By the Hurwitz theorem (see, for example, [55]) $\widetilde{h}$ is univalent on Δ.

Now again by Proposition 5.6.1 we have that $\widetilde{h}$ is a fanlike function the image of which lies in the wedge of angle π. Therefore the function h defined as

$$h(z) = \left[\widetilde{h}(z)\right]^{\mu/\beta}$$

is a univalent function on Δ whenever (5.7.23) holds. On account of (5.7.24) it is easy to see that h satisfies (5.7.22). $\square$

Remark 5.7.1 In the proof of the above proposition we have used an approximation process for the generator f (see formula 5.7.25). In turn, this process induces an approximation sequence of univalent functions $\widetilde{h}_n$ converging to $\widetilde{h}$ with interior null points defined by (5.7.27). However, in general we can claim only that these functions $\widetilde{h}_n$ are spirallike, but not necessarily starlike. In fact, we do not know whether the numbers μ_n in (5.7.27) are real.

5.8   Growth estimates for spirallike, starlike and fanlike functions

The famous Koebe distortion theorem asserts:

if $h \in S = \{h \in \mathrm{Univ}(\Delta) : h(0) = 0, \ h'(0) = 1\}$ then

$$\frac{|z|}{(1 + |z|)^2} \leq |h(z)| \leq \frac{|z|}{(1 - |z|)^2} \quad \text{for all } z \in \Delta.$$

Equality holds for the Koebe function

$$h_K(z) = \frac{z}{(1 - z)^2} = \sum_{n=1}^{\infty} n z^n.$$

Usually the proof of this fact is based on the known bound for the second coefficient in Taylor's expansion of $h \in S$, namely $|a_2| < 2$.

Note that the Bieberbach conjecture, namely that $|a_n| \leq n$ for each $h \in S$, $a_n = \frac{1}{n!}\frac{d^n h}{dz^n}(0)$, $n = 1, 2, \ldots$, was proved by L. de Branges [20] in his remarkable work in 1985, whilst for $h \in S^*(\Delta)$ it has been proved earlier by R. Nevanlinna [103].

Although analogous of the Koebe distortion theorem fail for biholomorphic mappings in the polidisks (or balls) of dimensions greater than 1, it is still relevant for many special cases, in particular, starlike and spirallike mappings.

Again as in previous sections one can use the relationships between these classes and semigroups to obtain corresponding bounds for these classes (see, for example, [109, 26]. Moreover, even in the one-dimensional case, by using some characteristics of semi-complete vector fields one can improve the estimates for some subclasses of Spiral(Δ).

1. We already know that if $h \in \mathrm{Hol}(\Delta, \mathbb{C})$ is a spirallike function on Δ, with

$$h(\tau) = 0, \quad h'(\tau) = \alpha \neq 0, \quad \tau \in \Delta,$$

then h satisfies the equation

$$\mu h(z) = h'(z) f(z) \qquad (5.8.1)$$

for some semi-complete vector field $f \in \mathrm{Hol}(\Delta, \mathbb{C})$ with $f(\tau) = 0$, $f'(\tau) = \mu \in \mathbb{C}$.

In addition, if $\{F_t\} = S_f$ is the semigroup generated by f, then h can be defined by the formula

$$h(z) = \alpha \lim_{t \to \infty} e^{\mu t} \cdot (F_t(z) - \tau) \qquad (5.8.2)$$

(see section 5.8).

First let us suppose that $\tau = 0$. It follows by Proposition 4.4.2 and Remark 4.4.4 that F_t satisfies the following estimate:

$$e^{-\mu t} \frac{|z|}{(1 + c|z|)^2} \leq \frac{|F_t(z)|}{(1 - c|F_t(z)|)^2} \leq e^{-\mu t} \frac{|z|}{(1 - c|z|)^2}$$

with some $c \in [0, 1]$. Moreover, if f is a bounded ρ-monotone function in Δ then c can be chosen strictly less then 1.

This inequality and (5.8.2) immediately imply the following estimates for h:

$$|h'(0)| \cdot \frac{|z|}{(1 + c|z|)^2} \leq |h(z)| \leq |h'(0)| \cdot \frac{|z|}{(1 - c|z|)^2}. \qquad (5.8.3)$$

For the general case, when $\tau \neq 0$, $\tau \in \Delta$ one can use the Möbius transformation M_τ:

$$M_\tau(z) = \frac{\tau - z}{1 - z\bar{\tau}},$$

defining the spirallike mapping $\widetilde{h} = h \circ M_\tau$ (note that the image $\widetilde{h}(\Delta) = h(\Delta)$).

Since $\widetilde{h}(0) = h(\tau) = 0$ and $\widetilde{h}'(0) = h'(\tau)(|\tau|^2 - 1)$ we obtain by (5.8.3) and replacing h by $\widetilde{h}$,

$$|h'(\tau)|(1 - |\tau|^2) \frac{|z|}{(1 + c|z|)^2} \leq |h(M_\tau(z))|$$

$$\leq |h'(\tau)|(1 - |\tau|^2) \frac{|z|}{(1 - c|z|)^2} \qquad (5.8.4)$$

or (replacing z by $M_\tau(z)$)

$$|h'(\tau)|(1 - |\tau|^2)\frac{d(\tau, z)}{(1 + cd(\tau, z))^2} \leq |h(z)|$$

$$\leq |h'(\tau)|(1 - |\tau|^2)\frac{d(\tau, z)}{(1 - cd(\tau, z))^2}, \qquad (5.8.5)$$

where $d(\tau, z) = |M_\tau(z)|$ is the pseudo-hyperbolic distance on Δ.

Thus we have proved the following assertion.

Proposition 5.8.1 *Let $h \in \mathrm{Hol}(\Delta, \mathbb{C})$ be a spirallike (starlike) function on Δ, with $h(\tau) = 0, \tau \in \Delta$.*

Then h satisfies estimates (5.8.5). Moreover, if h is strongly spirallike, then c can be chosen to be strictly less than 1.

Remark 5.8.1. To obtain a growth estimate for $h \in \mathrm{Spiral}(\Delta, \tau)$ in a form which does not contain a Möbius transformation we recall that the sets

$$D_\tau(K) = \left\{ z \in \Delta : \varphi_\tau(z) = \frac{|1 - z\bar\tau|^2}{1 - |z|^2} < K, \ K > 1 - |\tau|^2 \right\}$$

are F_t-invariant for all $t \geq 0$, and h belongs to $\mathrm{Spiral}(D_\tau(K), \tau)$ for each $K > 1 - |\tau|^2$. Therefore it may be convenient to rewrite the right hand side of (5.8.5) in terms of the function φ_τ. By simple calculations we obtain

$$\frac{1 - |\tau|^2}{(1 - c|M_\tau(z)|)^2} \leq \frac{(1 - |\tau|^2)(1 + c|M_\tau(z)|)^2}{(1 - |M_\tau(z)|^2)^2}$$

$$= \frac{(1 - |\tau|^2)(1 + c|M_\tau(z)|)^2}{(1 - |z|^2)^2(1 - |\tau|^2)^2} \cdot |1 - z\bar\tau|^4$$

$$= [\varphi_\tau(z)]^2 \frac{(1 + c|M_\tau(z)|)^2}{1 - |\tau|^2}.$$

Now (5.8.5) implies

$$|h(z)| \leq (1 + c)^2 [\varphi_\tau(z)]^2 \frac{|h'(0)|}{1 - |\tau|^2}. \qquad (5.8.6)$$

2. If h is a starlike function on Δ, normalized by the conditions

$$h(\tau) = 0, \quad h(0) = a \in \mathbb{C}, \qquad (5.8.7)$$

then one can find an estimate of growth by using Hummel's representation formula (see Section 5.5):

$$h(z) = H_\tau(z)g(z),$$

where

$$H_\tau(z) = \frac{(z - \tau)(1 - z\bar\tau)}{z}, \quad z \in \Delta,$$

and $g \in \text{Star}(\Delta, 0)$.

Indeed, it follows by the right hand inequality in (5.8.3) that

$$
\begin{aligned}
|h(z)| &= \frac{|z - \tau|\,|1 - z\bar{\tau}|}{|z|}|g(z)| \leq \frac{|z - \tau|\,|1 - z\bar{\tau}|}{(1 - |z|)^2} \cdot |g'(0)| \\
&= \varphi_\tau(z)\frac{1 + |z|}{1 - |z|}\frac{|z - \tau|}{|1 - z\bar{\tau}|} \cdot |g'(0)|.
\end{aligned}
$$

Similarly, by using the left hand inequality in (5.8.3) we obtain

$$
|h(z)| \geq \varphi_\tau(z)\frac{1 - |z|}{1 + |z|} \cdot \frac{|z - \tau|}{|1 - z\bar{\tau}|} \cdot |g'(0)|.
$$

Taking into account that

$$
g'(0) = -\tau a,
$$

we obtain the following estimate:

Proposition 5.8.2 *Let $h \in \text{Hol}(\Delta, \mathbb{C})$ be a starlike function on Δ normalized by conditions (5.8.7).*

Then the following estimate of growth holds:

$$
|h(0)|\,|\tau|\,d(\tau, z) \cdot \varphi_\tau(z)\frac{1 - |z|}{1 + |z|} \leq |h(z)|
$$
$$
\leq |h(0)|\,|\tau|\,d(\tau, z) \cdot \varphi_\tau(z)\frac{1 + |z|}{1 - |z|}. \qquad (5.8.8)
$$

3. Note that the latter arguments are relevant also when τ is a boundary of Δ. Moreover, in this case (5.8.8) can be written in a more precise form. Indeed, using Proposition 5.6.1(ii) we have that if h is a fanlike function on Δ, normalized by the conditions:

$$
h(1) = 0, \qquad h(0) = 1,
$$

then for some $\lambda \in [0, 1)$ it satisfies the equation

$$
h(z) = \frac{(1 - z)^{2 - 2\lambda}}{z}g(z),
$$

where $g \in S^*(\lambda)$ *(starlike of order λ)*, i.e.,

$$
\text{Re}\,\frac{zg'(z)}{g(z)} > \lambda.
$$

Using the latter inequality it is easy to see that

$$
\frac{g(z)}{z} = \left[\frac{g_1(z)}{z}\right]^{1 - \lambda},
$$

where g_1 belongs to S^* (cf., Section 5.6). Then again by (5.8.3) we obtain

$$\left[\frac{|1-z|}{(1+|z|)}\right]^{2-2\lambda} \le |h(z)| \le \left[\frac{|1-z|}{(1-|z|)}\right]^{2-2\lambda}. \tag{5.8.9}$$

In particular, we obtain that h is bounded in each nontangential approach region at $\tau = 1$. $\square$

Thus on account of Proposition 5.6.1 (see also Remark 5.6.1) and the obvious inequality $1 - |z| \le |1 - z| \le 1 + |z|$, we obtain from (5.8.9) the following distortion theorem.

Proposition 5.8.3 (see [134]) *Let h be a fanlike function on Δ normalized by the conditions*

$$h(1) = 0, \qquad h(0) = 1,$$

and let the image of h lie in the wedge of angle $\pi\nu$, $0 < \nu \le 2$.
Then the following estimates hold :

$$\left[\frac{1-|z|}{1+|z|}\right]^{\nu} \le |h(z)| \le \left[\frac{1+|z|}{1-|z|}\right]^{\nu}.$$

These estimates are sharp for the function $\left[\dfrac{1-z}{1+z}\right]^{\nu}$.

5.9   Remarks on Schroeder's equation and the Koenigs embedding property

The so called Schroeder's (functional) equation:

$$h(\varphi) = \lambda h \tag{5.9.1}$$

has been studied since the late nineteenth century (see, for example, [28] and references there).

Here φ is a given holomorphic self-mapping of Δ, where $\lambda \in \mathbb{C}$ is a suitable complex number, for which equation (5.9.1) has a solution $h \in \mathrm{Hol}(\Delta, \mathbb{C})$. Mostly this equation has been considered as an eigenvalue problem for a composition operator C_φ defined by the formula

$$C_\varphi h = h(\varphi) \tag{5.9.2}$$

on the space $\text{Hol}(\Delta, \mathbb{C})$ (or its relevant subspaces, in particular, Hardy spaces H^p).

In this section we shall discuss some relations of Schroeder's equation with spirallike and starlike mappings on Δ.

We assume that φ has an interior fixed point $a \in \Delta$. Of course, we will exclude the trivial cases when φ is a constant or the identity. In addition, we will assume that φ is not an elliptic automorphism of Δ.

In other words, our condition for φ is that $a \in \Delta$ is an attractive fixed point of it, or equivalently

$$|\varphi'(a)| < 1. \tag{5.9.3}$$

We will see below that in this case, in fact,

$$\varphi'(a) \neq 0 \tag{5.9.4}$$

is a necessary condition for the global solvability of (5.9.1).

First we note that if Schroeder's equation has a nontrivial solution (i.e., h is not constant), then λ is neither 0 nor 1.

Indeed, if $\lambda = 0$ then $h(\varphi(z)) = 0$ for $z \in \Delta$ and $h \equiv 0$ by the uniqueness theorem. If $\lambda = 1$ then h satisfied the equations

$$h(\varphi^{(n)}) = h, \quad n = 1, 2, \ldots,$$

and we have

$$h(z) = \lim_{n \to \infty} h(\varphi^{(n)}(z)) = h(a)$$

because of assumption (5.9.2). Thus we have to assume that λ is neither 0 nor 1.

Note that if $\lambda \neq 1$ then $h(a) = h(\varphi(a)) = \lambda h(a)$, and we have

$$h(a) = 0 \tag{5.9.5}$$

as a solution h of Schroeder's equation (5.9.1).

Hence h must have the form

$$h(z) = \sum_{n=k}^{\infty} a_n (z - a)^n, \quad k \geq 1. \tag{5.9.6}$$

For simplicity we put $a = 0$. Then (5.9.5) and (5.9.1) imply

$$\lambda = \frac{h(\varphi(z))}{h(z)} = \left(\frac{\varphi(z)}{z}\right)^k \frac{\displaystyle\sum_{n=k}^{\infty} a_n \varphi^{n-k}(z)}{\displaystyle\sum_{n=k}^{\infty} a_n z^{n-k}}.$$

Since the left hand side of this equality is a constant we obtain, letting z go to zero, that

$$\lambda = [\varphi'(0)]^k. \tag{5.9.7}$$

Since $\lambda \neq 0$ we obtain (5.9.3). $\square$

Thus we have proved the following assertion.

Proposition 5.9.1 *Let $\varphi \in \text{Hol}(\Delta)$ have an attractive fixed point $a \in \Delta$ and suppose that Schroeder's equation (5.9.1) has a nontrivial solution $h \in \text{Hol}(\Delta, \mathbb{C})$ for some $\lambda \in \mathbb{C}$. Then*

(i) $h(a) = 0$;

(ii) $\lambda = [\varphi'(0)]^k \neq 0$ for some positive k;

(iii) if h is locally univalent, then $\lambda = \varphi'(a) \neq 0$.

In 1884 Koenigs proved a remarkable result on the solvability of Schroeder's equation.

Proposition 5.9.2 (see [78]) *Let $\varphi \in \text{Hol}(\Delta)$ have an attractive fixed point $a \in \Delta$, such that*

$$\varphi'(a) := \lambda \neq 0.$$

Then there is a function $h \in \text{Hol}(\Delta, \mathbb{C})$ such that

$$h(\varphi(z)) = \lambda h(z)$$

for all $z \in \Delta$.

In addition, if φ is univalent then so is h.

The idea in the original proof of Koenigs's theorem is based on the convergence of the sequence

$$h_n(z) = \frac{\varphi^{(n)}(z)}{\lambda^n} = z + \sum_{k=2}^{\infty} a_k^{(n)} z^k, \quad n = 1, 2, \ldots,$$

which evidently satisfies the recursion equation

$$h_n(\varphi(z)) = \lambda h_{n+1}(z). \tag{5.9.8}$$

Its limit function h is called the *Koenigs function* and it is normalized by the condition

$$h'(a) = 1.$$

If such a function can be found one can present

$$\varphi(z) = h^{-1}(\lambda h(z)) \tag{5.9.9}$$

whenever the right hand side of this equality is well defined. Hence

$$\varphi^{(n)}(z) = h^{-1}(\lambda^n h(z)). \tag{5.9.10}$$

The latter expression then serves as a definition of fractional iterations of φ when n is not an integer (cf., section 4.4) and large enough.

Of course, if φ can be embedded in a globally defined continuous semigroup of holomorphic mappings it must be univalent, and so should h.

Although the discrete semigroup of iterates of φ cannot be embedded, in general, in a continuous semigroup of holomorphic self-mappings of Δ, depending on what one requires the answer may be yes in some suitable cases (see for example, G. Srekeres [138], J. Hadamard [59], T. Harris [63], C.C. Cowen [27]). The simplest case (though very useful) is, of course, when φ is a fractional linear transformation and so is h. We consider now, in some sense, the dual problem.

Definition 5.9.1 *We will say that a mapping* $\varphi \in \mathrm{Hol}(\Delta)$ *satisfies the Koenigs embedding property (K.e.p.) if its iterations* $\varphi^{(n)} : \Delta \rightarrow \Delta$ *can be embedded in a continuous semigroup* $\{F_t\}_{t \geq 0}$ *of holomorphic self-mappings of* Δ, *i.e.,* $F_1 = \varphi$.

It turns out that the answer to the question of what are the conditions for a univalent self-mapping $\varphi \in \mathrm{Hol}(\Delta)$ to satisfy the (K.e.p.) is related to some geometrical properties of the solution of Schroeder's equation.

Proposition 5.9.3 *Let* φ *be a univalent self-mapping of* Δ, *and let* $\varphi(\tau) = \tau$ *for some* $\tau \in \Delta$ *with* $0 < |\varphi'(\tau)| < 1$. *Then* φ *satisfies the Koenigs embedding property if and only if its Koenigs function is* μ-*spirallike, with* $\mu = -\log \varphi'(\tau)$.

Proof. Sufficiency. Suppose that equation (5.9.10) has a univalent μ-spirallike solution with $\mu = -\log \lambda$.

Then for each $z \in \Delta$ and $t \geq 0$ element $e^{-t\mu}h(z)$ belongs to $h(\Delta)$. If we define

$$F_t(z) = h^{-1}(e^{-t\mu}h(z)),$$

we obtain by (5.9.10) that

$$F_1(z) = h^{-1}(e^{\ln \lambda}h(z)) = h^{-1}(\lambda h(z)) = \varphi(z).$$

Necessity. Let φ satisfy the Koenigs embedding property, i.e., there is a semigroup $\{F_t\}_{t \geq 0} \subset \mathrm{Hol}(\Delta)$ such that

$$F_1(z) = \varphi(z).$$

Denote by $f = -\dfrac{dF}{dt}\big|_{t=0+}$ the generator of $\{F_t\}$ and consider the equation

$$\mu h(z) = f(z)\, h'(z), \tag{5.9.11}$$

where $\mu = -\log \lambda$.

Since $\mu = f'(\tau) \neq 0$ it follows by Proposition 5.7.2 that equation (5.9.11) has a univalent μ-spirallike solution h which satisfies the equality

$$h(F_t(z)) = e^{-\mu t}h(z).$$

Setting here $t = 1$ we obtain (5.9.10). $\square$

If φ has no fixed point in Δ then it follows by the Julia–Carathéodory Theorem that there is a unique point $\tau \in \partial\Delta$ such that $\angle \lim\limits_{z \to \tau} \varphi(z) = \tau$ and

$$0 < \angle \lim\limits_{z \to \tau} \varphi'(z) \leq 1.$$

Similarly, as in Proposition 5.9.3, by using Proposition 5.7.3 one proves the following assertion:

Proposition 5.9.4 *Let $\varphi \in \mathrm{Hol}(\Delta)$ have no fixed point in Δ, and let $\tau \in \partial\Delta$ be its sink point. Then φ satisfies the Koenigs embedding property if and only if for some $\lambda \in (0,1)$ there exists a solution of Schroeder's equation*

$$h(\varphi) = \lambda h, \qquad\qquad (5.9.12)$$

which is a univalent fanlike function on Δ such that $\angle \lim_{z \to \tau} h(z) = 0$.

Remark 5.9.1 In fact, Proposition 5.7.3 and the Julia–Wolff–Carathéodory Theorem imply that if φ satisfies the Koenigs embedding property then Shroeder's equation (5.8.11) is solvable if and only if $\alpha^2 \leq \lambda < 1$, where $\alpha = \angle \lim_{z \to \tau} \varphi'(z) > 0$.

Another direct consequence of the above propositions is a result originally established by A. Siskakis (see [136]).

Corollary 5.9.1 *Let F_t be a continuous semigroup of holomorphic self-mappings of Δ, and suppose that there is a point $\tau \in \overline{\Delta}$, such that $\lim_{t \to \infty} F_t(z) = \tau$, and $\angle \lim_{z \to \tau} \mathrm{Re}\, f'(z) \neq 0$, where $f = -\dfrac{dF}{dt}\big|_{t=0+}$.*
Then there are $\lambda \in \mathbb{C}$, $0 < |\lambda| < 1$, and $h \in \mathrm{Hol}(\Delta, \mathbb{C})$ such that for all $t \geq 0$

$$h(F_t) = \lambda h. \qquad\qquad (5.9.13)$$

In other words, there is a solution of Schroder's equation which does not depend on $t \geq 0$.

In addition, the solution of (5.9.12) can be found by solving the differential equations (5.9.11) with $\mu = -\log \lambda$.

Example 1. Let $\varphi(z) = \dfrac{z}{\sqrt{z^2 - e^2(z^2 - 1)}}$. It is easy to verify that φ satisfies Koenigs embedding property with $F_t(z) = \dfrac{z}{\sqrt{z^2 - e^{2t}(z^2 - 1)}}$, $F_t(0) = 0$, for all $t \geq 0$.

Therefore, instead of (5.9.10) one can try to solve (5.9.12) with $\lambda = \varphi'(0) = 1/e$ or (5.9.11) with $\mu = 1$.

Since $f(z) = -\dfrac{dF_t(z)}{dt}\big|_{t=0+} = z - z^3$ equation (5.9.11) becomes

$$h(z) = h'(z)(z - z^3).$$

Solving the latter equation we obtain that

$$h(z) = \frac{az}{\sqrt{1 - z^2}}, \qquad a \in \mathbb{C},$$

is a solution of equations (5.9.10) and (5.9.12).

The converse scheme does work if we know in advance the solution of Schroeder's equation. In particular, it can always be solved for a fractional linear transformation of the unit disk with an interior fixed point. Indeed, let $\varphi \in \mathrm{Hol}(\Delta)$ be a fractional linear mapping of Δ with a fixed point $\tau \in \Delta$. Without loss of generality we can consider the case $\tau = 0$. Then φ can be written in the form

$$\varphi(z) = \frac{az}{1 - bz}. \qquad (5.9.14)$$

We will investigate the sufficient and necessary conditions on coefficients a and b such that φ will satisfy the Koenigs embedding property.

First we note that the condition $|\varphi(z)| < 1$ for all $|z| < 1$ imply

$$|a| + |b| \le 1. \qquad (5.9.15)$$

In addition, we have $\varphi'(0) = a$.

Then it is easy to see that the function

$$h(z) = \frac{z}{1 - kz}, \qquad (5.9.16)$$

where $k = \dfrac{b}{1-a}$, is the Koenigs function for φ, i.e., $h(\varphi(z)) = \lambda h(z)$ with $\lambda = a$ and $h'(0) = 1$. Thus we tend to find a condition which will force h to be univalent μ-spirallike function with $\mu = -\log a$. In other words, we have to check the inequality

$$\operatorname{Re} \frac{\mu h(z)}{h'(z)} > 0. \qquad (5.9.17)$$

By (5.9.15) inequality (5.9.17) becomes

$$\operatorname{Re} \mu z(1 - kz) > 0, \qquad (5.9.18)$$

since by (5.9.15) $|k| \le 1$ the latter condition can be rewritten in the form

$$\cos \arg(\mu) \ge k. \qquad (5.9.19)$$

In particular, this condition always holds when μ is real.

Thus we have proved

Proposition 5.9.5 *Let $\varphi(z) = \dfrac{az}{1 - bz}$ with $|a| + |b| \le 1$, and $0 < |a| < 1$. Then φ satisfies the Koenigs embedding property if and only if*

$$\cos\left(\arg\left(\log \frac{1}{a}\right)\right) \ge \left|\frac{b}{1-a}\right|. \qquad (5.9.20)$$

In particular, if a is a real number then φ can be always embedded in a continuous semigroup of holomorphic self-mappings of Δ.

If we consider the mapping $\phi(z) = \dfrac{az}{1 - bz}$, as in Example 1 of Section 4.3, with $a = \frac{1}{3}\exp(i3.1)$ and $b = \frac{2}{3}$, then it is easy to see that $|a| + |b| = 1$ whilst condition (5.9.20) does not hold. Thus $\phi : \Delta \mapsto \Delta$ cannot be embedded in a continuous semigroup of **self-mappings** of Δ (see Figure 4.1).

Exercise 1. Prove the equivalence of conditions (5.9.17) and (5.9.18).

Exercise 2. Under condition (5.9.19) find the semigroup $F_t : \Delta \to \Delta$, $t \geq 0$, such that $F_1 = \varphi$, where φ has the form (5.9.14).

More detailed discussion of this approach for higher dimensions can be found in [76].

Because of our intention in a modest text to emphasize the dynamical flavor of the subject a big part of the theory which is primarily geometrical or functional analytic in nature has not been included.

We refer the reader to books of A.W. Goodman [57], P. Duren [33], and J.H. Shapiro [131], which could be good guides to complete the knowledge in these topics. Also, to advance to futher study on the boundary behavior of holomorphic mappings we mention an excellent book of C. Pommerenke [107].

Finally, we point out that many applicable subjects as branching processes, optimization theory, functional calculus etc., mentioned in the Preface may motivate an investigation in this direction.

Bibliography

[1] M. Abate, Horospheres and iterates of holomorphic maps, *Math. Z.* **198** (1988), 225–238.

[2] M. Abate, Converging semigroups of holomorphic maps, *Atti Accad. Naz. Lincei Rend. Cl. Sci. Fis. Mat. Natur.* **82** (1988), 223–227.

[3] M. Abate, *Iteration Theory of Holomorphic Maps on Taut Manifolds*, Mediterranean Press, 1989.

[4] M. Abate, Iteration theory, compactly divergent sequences and commuting holomorphic maps, *Ann. Scuola Norm. Sup. Pisa* **18** (1991), 167–191.

[5] M. Abate, The infinitesimal generators of semigroups of holomorphic maps, *Ann. Mat. Pura Appl.* **161** (1992), 167–180.

[6] D. Aharonov, M. Elin, S. Reich and D. Shoikhet, Parametric representations of semi-complete vector fields on the unit balls in $\mathbb{C}^n$ and Hilbert space, *Rend. Mat. Acc. Lincei*, **10** *4* (1999), 229–253.

[7] D. Aharonov, S. Reich and D. Shoikhet, Flow invariance conditions for holomorphic mappings in Banach spaces, *Math. Proceedings of the Royal Irish Academy*, **99A** (1999), 93–104.

[8] L. Aizenberg, S. Reich and D. Shoikhet, One-sided estimates for the existence of null points of holomorphic mappings in Banach spaces, *J. Math. Anal. Appl.* **203** (1996), 38–54.

[9] L. Aizenberg and A. Yuzhakov, *Integral representations and residues in multidimensional complex analysis* AMS, Providence, 1983.

[10] J.W. Alexander, Functions which map the interior of the unit circle upon simple regions, *Annals of Math.* **17** (1915), 12–22.

[11] I.A. Aleksandrov, *Parametric Extensions in the Theory of Univalent Functions*, Nauka, Moscow, 1976.

[12] J. Arazy, An application of infinite dimensional holomorphy to the geometry of Banach spaces, *Lecture Notes in Math.* **1267** (1987), 122–150.

[13] V. Barbu, *Nonlinear Semigroups and Differential Equations in Banach Spaces*, Noordhoff, Leyden, 1976.

[14] R.W. Barnard, C.H. FitzGerald and S. Gong, The growth and 1/4-theorems for starlike mappings in $\mathbb{C}^n$, *Pacific J. Math.* **150** (1991), 13–22.

[15] D.F. Behan, Commuting Analytic Functions Without Fixed Points, *Proc. of the Americ. Math. Society* **37** *1* (1973), 114–120.

[16] E. Berkson, R. Kaufman and H. Porta, Möbius transformations on the unit disk and one-parametric groups of isometries of H^p, *Transections Amer. Mathem. Soc.* **199** (1974), 223–239.

[17] E. Berkson and H. Porta, Semigroups of analytic functions and composition operators, *Michigan Math. J.* **25** (1978), 101–115.

[18] S.D. Bernardi, *Bibliography of Schlicht Functions*, Mariner Publ. Co., Tampa, FL, 1983.

[19] H. Brézis, *Operateurs Maximaux Monotenes*, North Holland, Amsterdam, 1973.

[20] L. Brickman, Φ-like analytic functions, I, *Bull. Amer. Math. Soc.* **79** (1973), 555–558.

[21] C. Carathéodory, Untersuchungen über die konformen Abbildungen von festen und veränderlichen Gebiten, *Math. Ann.*, **72** (1912), 107–144.

[22] C. Carathéodory, Uber die Winkelderivierten von beschränkten Analytischen Funktionen, *Sitzungsber.Preuss. Acad.Viss. Berlin*, Phys.-Math. Kl.(1929), 39–54.

[23] C. Carathéodory, *Theory of functions of a complex variable*, Chelsea, 1954.

[24] G.-N. Chen, Iteration for Holomorphic Maps of the Open Unit Ball and the Generalized Upper Half-Plane of $\mathbb{C}^n$, *Journ. of Mathemat. Analysis and Appl.* **98** (1984), 305–313.

[25] P. Chernoff and J.E. Marsden, On continuity and smoothness of group actions, *Bull. Amer. Math. Soc.* **76** (1970), 1044–1049.

[26] M. Chuaqui, Applications of subordination chains to starlike mappings in $\mathbb{C}^n$, *Pacific J. Math.*, **168** (1995), 33–48.

[27] C.C. Cowen, Iteration and the solution of functional equations for functions analytic in the unit disk, *Trans. Amer. Math. Soc.* **265** (1981), 69–95.

[28] C.C. Cowen and B.D. MacCluer, *Composition Operators on Spaces of Analytic Functions*, CRC Press, Boca Raton, FL, 1995.

[29] M.G. Crandall and T.M. Ligett, Generation of semigroups of nonlinear transformations on a general Banach space, *Amer. J. Math.* **93** (1971), 265–298.

[30] L. de Branges, A proof of the Bieberbach conjecture, *Acta Math.* **154** (1985), 137–152.

[31] A. Denjoy, Sur l'itération des fonctions analytiques, *C. R. Acad. Scie.* **182** (1926), 255–257.

[32] S. Dineen, *The Schwartz Lemma*, Clarendon Press, Oxford, 1989.

[33] P. Duren, *Univalent Functions*, Springer, 1983.

[34] C.J. Earle and R. S. Hamilton, A fixed point theorem for holomorphic mappings, *Proc. Symp. Pure Math.*, Vol. 16, Amer. Math. Soc., Providence, R.I. (1970), 61–65.

[35] E. Egervàry, Abbildungseigenschaften der arithmetischen Mittel der geometrischen Reihe, *Math. Z.* **42** (1937), 221–230.

[36] M. Elin, S. Reich and D. Shoikhet, Complex Dynamical Systems and the Geometry of Domains in Banach Spaces, Technion preprint series.

[37] M. Elin, S. Reich and D. Shoikhet, Dynamics of Inequalities in Geometric Function Theory, *Journ. of Inequalities and Applications* (excepted for publication).

[38] M. Elin, S. Reich and D. Shoikhet, Holomorphically Accretive Mappings and Spiral-shaped Functions of Proper Contractions, *Nonlinear Analysis Forum* **5** (2000) 149–161.

[39] M. Elin, S. Reich and D. Shoikhet, Asymptotic behavior of semigroups of ρ-nonexpansive and holomorphic mappings on the Hilbert ball, Technion preprint series, (2000).

[40] M. Elin, S. Reich and D. Shoikhet, Asymptotic behavior of semigroups of holomorphic mappings, *Progress in Nonlinear Differential Equations and Their Applications*, Birkhauser Verlag, Basel, **42** (2000), 249–258.

[41] M. Elin, S. Reich and D. Shoikhet, A semigroup approach to the geometry of domains in complex Banach spaces, *Nonlinear Analysis*, accepted for publication.

[42] M. Elin and D. Shoikhet, Dynamic extension of the Julia-Wolff-Carathéodory Theorem, *Dynamic Systems and Applications* (accepted for publication).

[43] Ky Fan, Julia's lemma for operators, *Math. Ann.* **239** (1979), 241–245.

[44] Ky Fan, Iterations of analytic functions of operators, *Math. Z.* **179** (1982), 293–298.

[45] Ky Fan, Iterations of analytic functions of operators II, *Linear and Multtilinear Algebra* **12** (1983), 295–304.

[46] F. Forelli, The isometries of H^p, *Canad. J. Math.* **16** (1964), 721–728. MR 29 #6336.

[47] T. Franzoni and E. Vesentini, *Holomorphic Maps and Invariant Distances*, North-Holland, Amsterdam, 1980.

[48] I.I. Gikhman and A.V. Skorokhod, *Theory of Random Processes*, Nauka, Moskow, 1973.

[49] I. Glicksberg, Julia's lemma for function algebras, *Duke Mathematical Journal* **43** *2* (1976).

[50] K. Goebel, Fixed points and invariant domains of holomorphic mappings of the Hilbert ball, *Nonlinear Analysis* **6** (1982), 1327–1334.

[51] K. Goebel and S. Reich, Iterating holomorphic self-mappings of the Hilbert ball, *Proc. Japan Acad.* **58** (1982), 349–352.

[52] K. Goebel and S. Reich, *Uniform Convexity, Hyperbolic Geometry and Nonexpansive Mappings*, Marcel Dekker, New York and Basel, 1984.

[53] K. Goebel, T. Sekowski and A. Stachura, Uniform convexity of the hyperbolic metric and fixed points of holomorphic mappings in the Hilbert ball, *Nonlinear Analysis* **4** (1980), 1011–1021.

[54] J.L. Goldberg, Functions with positive real part in a half-plane, *Duke Math. J.* **29** (1962), 335–339.

[55] G.M. Golusin, *Geometric Theory of Functions of a Complex Variable*, Amer. Math. Soc., Providence, RI, 1969.

[56] S. Gong, *Convex and starlike mappings in several complex variables*, Science Press, Beijing–New York & Kluwer Acad. Publ., Dordrecht–Boston–London.

[57] A.W. Goodman, *Univalent Functions*, Vols. I, II, Mariner Publ. Co., Tampa, FL, 1983.

[58] K.R. Gurganus, Φ-like holomorphic functions in $\mathbb{C}^n$ and Banach space, *Trans. Amer. Math. Soc.* **205** (1975), 389–406.

[59] J. Hadamard, Two works on iteration and related questions, *Bull. Amer. Math. Soc.* **50** (1944), 67–75.

[60] L.A. Harris, The numerical range of functions and best approximation, *Proc. Camb. Phil. Soc.* **76** (1974), 133–141.

[61] L.A. Harris, On the size of balls covered by analytic transformations, *Monatshefte für Mathematik* **83** (1977), 9–23.

[62] L.A. Harris, S. Reich and D. Shoikhet, Dissipative holomorphic functions, Bloch radii, and the Schwarz Lemma, *J. Analyse Math.* **82** (2000), 221–232.

[63] T.E. Harris, *The Theory of Branching Processes*, Springer, Berlin, 1963.

[64] U. Helmke and J.B. Moore, *Optimization and Dynamical Systems*, Springer, London, 1994.

[65] J.A. Hummel, Multivalent starlike functions, *J. Analyse Math.* **18** (1967), 133–160.

[66] J.A. Hummel, Extremal properties of weakly starlike p-valent functions, *Trans. Amer. Math. Soc.* **130** *3* (1968), 544–551.

[67] J.A. Hummel, The coefficients of starlike functions, *Proc. Amer. Math. Soc.* **22** (1969), 311–315.

[68] A. Hurwitz and R. Courant, *Funktionentheorie*, Springer Verlag, Berlin, 1929.

[69] M.E. Jacobson, Computation of extinction probabilities for the Bellman–Harris branching process, *Math. Biosciences* **77** (1985), 173–177.

[70] F. Jafari and K. Yale, Cocycles, coboundaries and spectra of compositions operators, *Proc. of the Second Big Sky Conf. in Analysis*, Eastern Montana College, Billings, MT (1994).

[71] G. Julia, Extension nouvelle d'un lemme de Schwarz, *Acta Math.* **42** (1920), 349–355.

[72] G. Julia, *Principes géométriques d'analyse*, Gauthier–Villars, Paris, Part I (1930), Part II (1932).

[73] W. Kaplan, Close-to-convex schlicht functions. *Michigan Math. J.* **1** (1952), 169–185.

[74] S. Karlin and J. McGregor, Embeddability of discrete time simple branching processes into continuous time processes, *Trans. Amer. Math. Soc.* **132** (1968), 115–136.

[75] V. Khatskevich, S. Reich and D. Shoikhet, Complex dynamical systems on bounded symmetric domains, *Electronic J. Differential Equations* **19** (1997), 1–9.

[76] V. Khatskevich, S. Reich and D. Shoikhet, Schröder's functional equation and the Koenigs embedding property, *Nonlinear Analysis*, accepted for publication.

[77] V. Khatskevich and D. Shoikhet, *Differentiable Operators and Nonlinear Equations*, Birkhäuser, Basel, 1994.

[78] G. Koenigs, Recherches sur les intégrales de certaines équations fonctionnelles, *Ann. Sci. École Norm. Sup.* **1** (1884), 2–41.

[79] M.A. Krasnoselskii and P.P. Zabreiko, *Geometrical Methods of Nonlinear Analysis*, Springer, Berlin, 1984.

[80] Y. Kubota, Iteration of Holomorphic Maps of the unit Ball into itself, *American Mathem. Society*, **88** it 3 (1983).

[81] T. Kuczumow, S.Reich and D.Shoikhet, The existence and non-existence of common fixed points for commuting families of holomorphic mappings, *Nonlinear Anal.*, **43** (2001), 233–251.

[82] T. Kuczumow and A. Stachura, Convexity and fixed points of holomorphic mappings in Hilbert ball and polydisc, *Bull. Acad. Polon. Sci.* **34** (1986), 189–193.

[83] T. Kuczumow and A. Stachura, Common fixed points of commuting holomorphic mappings, *Kodai J. Math.* **12** (1989), 423–428.

[84] T. Kuczumow and A. Stachura, Iterates of holomorphic and k_D-nonexpansive mappings in convex domains in $\mathbb{C}^n$, *Advances in Math.* **81** (1990), 90–98.

[85] Y.D. Latuskin and A.M. Stepin, Weighted shift operators and linear extensions of dynamical systems, *Uspekhi Mat. Nauk* **46** (1991), 85–143.

[86] E. Landau and G. Valiron, A deduction from Schwarz's Lemma, *J.London Math. Soc.* **4** (1929), 162–163.

[87] N.A. Lebedev, *The square principle in the Theory of Univalent Functions*, Nauka, Moskow, 1975.

[88] E. Lindelöf, Mémoire sur sertaines inégalités dans la théorie des fonctions monogénes et sur quelques properiétés nouvelles de ces fonctions dans le voisinage d'un point singulier essentiel, *Acta Soc. Sci. Fennicae* **35** 7 (1909).

[89] K. Löwner, Untersuchungen über schlichte konforme Abbildungen des Einheitskreises, I, *Math. Ann.* **89** (1923), 103–121.

[90] A. Lyzzaik, On a conjecture of M.S. Robertson, *Proc. Amer. Math. Soc.* **91** (1984), 108–110.

[91] B.D. MacCluer, Iterates of Holomorphic Self-Maps of the unit Ball in $\mathbb{C}^n$, *Michigan Math. J.* **30** (1983).

[92] P. Mazet and J.P. Vigué, Points fixes d'une application holomorphe d'un domaine borné dans lui-même, *Acta Math.* **166** (1991), 1–26.

[93] P. Mazet and J.P. Vigué, Convexité de la distance de Carathéodory et points fixes d'applications holomorphes, *Bull. Sc. Math.* **16** (1992), 285–305.

[94] P. Mellon, Another look at results of Wolff and Julia type for J^*-algebras, *J.Math. Anal. Appl.* **198** (1996), 444–457.

[95] P.R. Mercer, Another look at Julia's Lemma, Preprint, 1991.

[96] P.R. Mercer, Proper Maps, Complex Geodesics and Iterates of Holomorphic Maps on Convex Domains in $\mathbb{C}$, *Contemporary Mathematics*, **137** (1992).

[97] P.R. Mercer, Sharpened Versions of the Schwarz Lemma, *J. of Math. Anal. and Appl.* **205** (1997), 508–511 Article No. AY975217.

[98] P.R. Mercer, Note on a strengthened Schwarz–Pick inequality, *Journ. Math. Anal. Appl.* **234** (1999), 735–739.

[99] P. Montel, Sur les familles normales de fonctions analytiques, *Ann. Sci. École Norm. Sup.* **33** 3 (1916), 223–302.

[100] P. Montel, *Leçons sur les familles normales de fonctions analytiques et leurs applications*, Gauthier–Villars, Paris, 1927.

[101] P. Montel, *Leçons sur les Fonctions Univalentes ou Multivalentes*, Gauthier–Villars, Paris, 1933.

[102] R. Nevanlinna, Über die konforme Abbildung Sterngebieten, *Oeversikt av Finska-Vetenskaps Societeten Ferhandlinger* **63(A)**, No. 6, 1921.

[103] R. Nevanlinna, *Analytic Functions*, Springer, New York, 1970.

[104] J.A. Pfaltzgraff, Subordination chains and univalence of holomorphic mappings in $\mathbb{C}^n$, *Math. Ann.* **210** (1974), 55–68.

[105] J.A. Pfaltzgraff, Subordination chains and quasiconformal extension of holomorphic maps in $\mathbb{C}^n$, *Ann. Acad. Sci. Fen.* Ser. A Math. **1** (1975), 13–25.

[106] J.A. Pfaltzgraff and T.J. Suffridge, Close-to-starlike holomorphic functions of several variables, *Pacific J. Math.* **57** (1975), 271–279.

[107] Ch. Pommerenke, *Univalent Functions*, Vandenhoech and Ruprecht, Göttingen, 1975.

[108] Ch. Pommerenke, On the iteration of analytic functions in a half plane I, *J. London Mat. Soc.* **19** (1979), 439–447.

[109] T. Poreda, On generalized differential equations in Banach space, Dissert. Math. Vol. **310**, Warsaw, 1991.

[110] V.P. Potapov, The multiplicative structure of J-contractive matrix functions, *Amer. Math. Soc. Transl.* **2** *15* (1960), 231–243.

[111] S. Reich, Averaged mappings in the Hilbert ball, *J. Math. Anal. Appl.*, **109** (1985), 199–206.

[112] S. Reich, Nonliear semigroups, holomorphic mappings, and integral equations, in *Proc. Symp. Pure Math.*, **vol. 45, Part II** Amer. Math. Soc., Providence, RI (1986), 307–324.

[113] S. Reich, The asymptotic behavior of a class of nonlinear semigroups in the Hilbert ball, *J. Math. Anal. Appl.* **157** (1991), 237–242.

[114] S. Reich and I. Shafrir, Nonexpansive iterations in hyperbolic spaces, *Nonlinear Analysis* **15** (1990), 137-158.

[115] S. Reich and D. Shoikhet, Generation theory for semigroups of holomorphic mappings in Banach spaces, *Abstr. Appl. Anal.* **1** (1996), 1–44.

[116] S. Reich and D. Shoikhet, Semigroups and generators on convex domains with the hyperbolic metric, *Atti. Acad. Naz. Lincei* (9) **8** (1997), 231–250.

[117] S. Reich and D. Shoikhet, The Denjoy–Wolff theorem, *Ann. Univ. Mariae Curie–Sklodowska* **51** (1997), 219–240.

[118] S. Reich and D. Shoikhet, Metric domains, holomorphic mappings and non-linear semigroups, *Abstr. Appl. Anal.* **3** (1998), 203–228.

[119] M.S. Robertson, On the theory of univalent functions, *Ann. of Math.*, **37** (1936), 374–408.

[120] M.S. Robertson, Applications of the subordination principle to univalent functions, *Pacific J. Math.* **11** (1961), 315–324.

[121] M.S. Robertson, Univalent functions starlike with respect to a boundary point, *J. Math. Anal. Appl.* **81** (1981), 327–345.

[122] W. Rudin, *Real and Complex Analysis*, McGraw-Hill, New York, 1974.

[123] D. Sarason, Angular derivatives via Hilbert space, *Complex Variables*, **10** (1988), 1–10.

[124] E. Schroeder, Über itierte Funktionen, *Math. Ann.* **3** (1871), 296–322.

[125] L. Schwartz, *Analyse Mathematique*, Hermann, 1967.

[126] J. Serrin, A note on harmonic functions defined in a half-plane, *Duke Math. J.* **23** (1956), 523–526.

[127] B.A. Sevastyanov, *Branching Processes*, Nauka, Moscow, 1971.

[128] B.V. Shabat, *Introduction to complex analysis*, Nauka, Moscow, 1976.

[129] I. Shafrir, Common fixed points of commuting holomorphic mappings in the product of n Hilbert balls, *Michigan Math. J.* **39** (1991), 281–287.

[130] I. Shafrir, Coaccretive operators and firmly nonexpansive mappings in the Hilbert ball, *Nonlinear Analysis* **18** (1992), 637–648.

[131] J.H. Shapiro, *Composition Operators and Classical Function Theory*, Springer, Berlin, 1993.

[132] A.L. Shields, On fixed points of commuting analytic functions, May 16, 1963.

[133] D. Shoikhet, Some properties of Fredholm mappings in Banach analytic manifolds, *Integral Equations Operator Theory* **16** (1993), 430–451.

[134] H. Silverman and E.M. Silvia, Subclasses of univalent functions starlike with respect to a boundary point, *Houston J. Math.* **16** (1990), 289–299.

[135] R. Sine, Behavior of iterates in the Poincaré metric, *Houston Journal of Mathematics* **15** *2* (1989).

[136] A. Siskakis, Composition semigroups and the Cesaro operator on H^p, *Journ. London Math. Soc.* **36** (1987), 153–164.

[137] L. Špaček, Přispěvek k teorii funkci prostych, *Časopis Pěst. Mat.*, **62** (1933), 12–19.

[138] G. Srekeres, Regular iteration of real and complex functions, *Acta Math.* **100** (1958), 203–258.

[139] J.F. Steffensen, Om sandsynligheden for at afkommet uddor, *Matematisk Tidsskrift* B1 (1930), 19–23.

[140] T.J. Suffridge, Starlike and convex maps in Banach space, *Pacific J. Math.* **46** (1973), 575–589.

[141] T.J. Suffridge, Starlikeness, convexity and other geometric properties of holomorphic maps in higher dimensions, *Complex Analysis (Proc. Conf. Univ. Kentucky, Lexington, KY, 1976)*, Lecture Notes in Math. **599** (1977), 146–159.

[142] H. Upmeier, *Jordan Algebras in Analysis, Operator Theory and Quantum Mechanics*, CBMS, Reg. Conf. Ser. in Math., Vol. **67**, Amer. Math. Soc., Providence, RI, 1986.

[143] G. Valiron, Sur l'iteration des fonctions holomorphes dans un demi-plan, *Bull. des Sci. Math.* **55** *2* (1931), 105–128.

[144] E. Vesentini, Su un teorema di Wolff e Denjoy, *Rend. Sem. Mat. Fis. Milano* **53** (1983), 17–25.

[145] E. Vesentini, Iterates of holomorphic mappings, *Uspekhi Mat. Nauk* **40** (1985), 13–16.

[146] E. Vesentini, Semigroups of holomorphic isometries, *Advances in Math.* **65** (1987), 272–306.

[147] E. Vesentini, Krein spaces and holomorphic isometries of Cartan domains, *Geometry and Complex Variables*, Marcel Dekker, New York, 1991, 409–413.

[148] E. Vesentini, Semigroups of linear contractions for an indefinite metric, *Met. Mat. ACC. Lincei* **9** *2* (1994),53–83.

[149] G. Vitali, Sopra le serie di funzione analitiche, *Rend. del R. Instituto Lombardo di Scienze e Lettere* **36** *2* (1903), 772–774.

[150] J.K. Wald, *On starlike functions*, (Ph.D. Thesis), Univ. of Delaware, Newark, Delaware, 1978.

[151] H.W. Watson and F. Galton, On the probability of the exinction of families, *Jour. Anihrop. Inst.* **4** (1874), 138–144.

[152] K. Wlodarczyk, Iterations of Holomorphic Maps of Infinite Dimensional Homogeneous Domains, *Mh. Math.* **99** (1985), 153–160.

[153] K. Wlodarczyk, *Studies of iterations of Holomorphic maps in j-Algebras and complex Hilbert Spaces* Oxford University Press, 1986.

[154] K. Wlodarczyk, Julia's Lemma and Wolff's Theorem for j-Algebras, *American Math. Society* **99** *3* March 1987.

[155] J. Wolff, Sur l'iteration des fonctions holomorphes dans une region, et dont les valeurs appartiennent a cette region, *C. R. Acad. Sci.* **182** (1926), 42–43.

[156] J. Wolff, Sur l'iteration des fonctions bornees, *C. R. Acad. Sci.* **182** (1926), 200–201.

[157] J. Wolff, Sur une generalisation d'un theoreme de Schwarz, *C. R. Acad. Sci.* **182** (1926), 918–920.

[158] J. Wolff, L'équation differetielle $dz/dt = w(z)$ =fonction holomorphe à partie réelle positive dans un demi-plan, *Compos. Math.* **6** (1938), 296–304.

[159] K. Yosida, *Functional Analysis*, Springer, Berlin, 1968.

[160] P. Yang, Holomorpic curves and boundary regularity of biholomorpic maps of pseudoconvex domains, *Preprint* (1978).

Author and Subject Index

Abate, M., vii, 101, 205

Alexander, J. W., 158, 205

Behan, D.F., 37, 206

Berkson, E., vii, 63, 93, 95, 101, 206

Bernardi, S.D., viii, 206

Brickman, L., ix, 206

Carathéodory, C., 9, 17, 206

Chen, G.N., vii, 206

Cowen, C.C., viii, 200, 207

de Branges, L., ix, 194, 207

Denjoy, A., vi, 9, 32, 35, 207

Earle, C., vii, 207

Egervàry, E., 176, 207

Fan, K., vii, 208

Forelli, F., vii, 208

Galton, F., v, 215

Glicksberg, I., vii, 208

Goebel, K., vii, 208

Gurganus, K.R., ix, 209

Hadamard, J., 200, 209

Hamilton, R., vii, 207

Harris, L.A., 13, 209

Harris, T.E., vi, 200, 209

Herglotz, G., 5

Hummel, J.A., 171, 172, 174, 209

Jafari, F., vii, 209

Julia, G., vii, 9, 17, 209

Koenigs, G., vi, 200, 210

Kubota, Y., vii, 210

Kuczumov, T., vii, 210

Löwner, K., ix, 211

Latuskin, Y., vii, 210

Lyzzaik, A., viii, 171, 177, 211

MacCluer, B., vii, viii, 207, 211

Mazet, P., vii, 211

Mellon, P., vii, 211

Mercer, P., vii, 211

Montel, P., 6, 211

Nevanlinna, R., viii, 94, 158, 194, 212

Pfaltzgraff, J.A., ix, 212

Pick, G., 9

Poincaré, H., 7, 43, 44

Porta, H., vii, 63, 93, 95, 101, 206

Potapov, V.P., vii, 212

Reich, S., vii, 205, 207–210, 212

Riemann, B., 7

Riesz, F., 5

Robertson, M.S., viii, ix, 159, 171, 176, 213

Schroeder, E., 198, 213

Schwarz, H.A., 9

Sekowski, T., vii, 208

Sevastyanov, B.A., vi, 213

Shafrir, I., vii, 213

Shields, A.L., 36, 213

Silverman, H., viii, 177, 179, 213

Silvia, E.M., viii, 177, 179, 213

Siskakis, A.G., vii, 214

Srekeres, G., 200, 214

Stachura, A., vii, 208, 210

Steffensen, J.F., vi, 214

Stepin, M., vii, 210

Suffridge, T.J., ix, 212, 214
Vesentini, E., vii, 101, 208, 214
Vigué, J.P., vii, 211
Wald, J.K., viii, 159, 215
Watson, H.W., v, 215
Wlodarczyk, K., vii, 215
Wolff, J., vi, vii, 9, 32, 35, 101, 215
Yale, K., vii, 209

admissible curve, 42, 45–47
angular derivative, 23, 32, 35, 135, 140, 145, 150, 164
angular limit, 20, 23, 186
approximating curve, vii, 53, 55, 56
asymptotic behavior, vii, 83, 98, 103, 113, 135, 142–145, 150, 177
automorphism, vi, 9, 15, 19, 26, 30, 35, 56, 83, 146
    elliptic, 27, 30, 36, 103, 105, 112, 113
    hyperbolic, 27–29, 32, 37, 105, 109, 111
    parabolic, 27, 28, 30, 32, 105, 111, 112

Banach Fixed Point Principle, 8, 31, 52, 99
Berkson–Porta representation, 95, 98, 99, 139, 162, 170, 171, 174, 181
Bieberbach conjecture, ix, 194
Bohl–Poincaré theorem, 81
boundary behavior, 17, 163
branching process, v, vi
Brouwer's Fixed Point Principle, 8, 31

Carathéodory Kernel Theorem, viii, 6, 166, 177, 183
Cauchy Integral Formula, 4

Cauchy problem, 66, 68, 69, 72, 76, 82, 83, 87, 103, 161
Cauchy–Schwarz formula, 5
class of Carathéodory, 89, 178
commuting family, 36, 101
complete metric space, 7, 43, 44
complete vector field, 104, 156
contraction, 7, 17
    strict, 8, 31, 52, 70, 99, 122

Denjoy–Wolff Theorem, vii, 30, 32, 36, 101, 102

embedding property, 109, Chapter V
Euclidean distance, 7, 13, 114
exponential formula, 68, 70, 97, 122

fixed point, 8, 27, 31, 32, 53, 54, 57, 99, 111
    attractive, 32, 35, 156, 199, 200
    boundary, 26, 29, 30, 32, 35
    common, 36, 101
    free mapping, 33, 37, 54
    interior, vi, vii, 25, 30–32, 56
    of automorphisms, 26
    of holomorphic self-mapping, 25, 31
    of nonexpansive mapping, 52
flow invariance condition, vii, 82, 98
fractional linear transformation, 10, 29, 40, 107, 109, 200, 202
function
    ρ-monotone, 79–82, 98, 141, 195
        strongly, 98, 99, 169
    close-to-convex, ix, 153, 177
    fanlike, 157, 176, 192, 198
    harmonic, 5, 85, 86, 97, 175
    holomorphic, 4, 9
    spirallike, see spirallike function
    starlike, see starlike function

univalent, ix, 4, 6, 153, 154, 158, 159, 161, 162, 170, 177, 178, 180

generator, 66, 68, 81, 82, 144, 147, 150, 160, 162
  of a one-parameter group, 82, 83, 104
  of a one-parameter semigroup, 70, 76, 83
geodesic segment, 15, 45, 46
geodesics, 44, 47
group
  of automorphisms, 82, 104
  one-parameter, 60
growth estimate, 194, 196

Harnack inequality, 89, 146, 149
  strong, 90, 119
horocycle, 18, 19, 32, 37, 54, 97, 136, 167
Hummel's multiplier, 171, 172, 177
Hurwitz convergence theorem, 6
hyperbolic
  ball, 136, 174
  distance, 45
  length, 39, 42, 45
  metric, vii, 7, 8, 39, 72, 79, 98, 99, 122

Implicit Function Theorem, 72, 82
infinitesimal generator, 66, 70, 76, 83
involution property, 11
isometry, ρ-isometry, 52, 60

Julia number, 19, 33, 35
Julia's Lemma, vi, 18, 19, 24, 32, 33, 35, 151
Julia–Carathéodory Theorem, 22, 23, 32, 35, 150, 201

Julia–Wolff–Carathéodory Theorem, 33, 164, 171, 202

Koebe distortion theorem, 182, 194, 195
Koebe function, 168, 194
Koebe One-Quarter Theorem, 153
Koenigs embedding property, 198, 201 203
Koenigs function, 200, 201

Lebesgue's Bounded Convergence Theorem, 185
Lindelöf's inequality, 12
Lindelöf's Principle, 23, 182
lower bound, 80, 135, 144
  appropriate, 137, 140, 142, 143, 148

Möbius transformation, vii, 10, 11, 15, 40, 46, 83, 168, 172, 173, 195, 196
mapping
  conformal at the boundary point, 164
  identity mapping, 11
  isogonal at the boundary point, 164
  nonexpansive, 7, 8, 39, 43
    ρ-nonexpansive, 52, 56, 57, 68–70, 75–77, 79, 81, 82, 136, 144
  fixed point free, 54
  fixed point of, 52
maximum principle
  for harmonic functions, 5, 85, 86, 96
  maximum modulus principle, 5, 10, 35, 47
Montel Theorem, 6

Nevanlinna's condition, ix, 163, 167
nontangential approach region, 20, 23,
    150, 182, 198
nontangential limit, 20, 21, 34
normal family, 6
null point, 98, 104, 135, 144, 156–158,
    161, 171

Open Mapping Theorem, 154

Poincaré metric, 7, 8, 39, 43–46, 70,
    72, 79, 98
    infinitesimal, 44, 45, 122
power convergence, 32, 35, 36, 56, 57
pseudo-hyperbolic
    ball, 17, 25, 118, 167
    disk, 32
    distance, 14, 17, 39, 114, 196
    metric, 7

range condition, 69, 70, 72, 74, 77,
    79, 82, 88
rate of convergence, 29, 35, 98, 113,
    135, 139, 142, 144, 147, 149
resolvent, 73–78, 81, 82, 96, 99, 193
    nonlinear, 67, 69, 72, 79, 82
resolvent identity, 74–78
retraction, 56, 78
Riemann Mapping Theorem, viii, 154,
    166
Riesz–Herglotz representation, 5, 93,
    178, 184

Schroeder's equation, 198–201
Schwarz Lemma, vi, 9, 12, 27, 87, 89,
    91, 149, 167, 188
Schwarz–Pick inequality, 11, 13, 17,
    19, 43, 82
Schwarz–Pick Lemma, 11, 14, 17, 19,
    25, 26, 31, 32, 57
    boundary version, 35

semi-complete vector field, 91, 95, 156,
    174, 192, 195
semigroup, ix, 9, 60, 68, 97, 98, 135,
    150, 155, 187, 190
    one-parameter, v, 70
    continuous, 60, 63, 66, 109
    discrete, 26, 60, 109
sink point, 33, 35–37, 55–57, 78, 97,
    105, 110–112, 135, 139, 142,
    144, 149, 150, 171, 182, 185,
    192
spirallike function, viii, ix, 159, 160,
    167, 190, 191, 195
    strongly, 169, 196
    with respect to a boundary point,
        159, 162
    with respect to an interior point,
        159, 161, 162
spiralshaped set, 153, 159, 160, 166,
    167
starlike function, viii, ix, 157, 158,
    160, 162, 167, 169, 170, 190,
    191, 196
    of order λ, 159, 179, 197
    with respect to a boundary point,
        viii, 157, 158, 171, 176, 177,
        180, 185, 192
    with respect to an interior point,
        157, 159, 161, 170, 171, 176,
        177
starshaped set, 153, 157, 159–161, 166,
    167, 172, 178
stationary point, 101, 104, 136, 138,
    142
Stolz angle, 21
strict contraction, 8, 31, 52, 70, 99,
    122
support functional, 136

Taylor representation, 5, 10, 154, 194

uniform convergence, 6, 54, 142
uniform Lipschitz condition, 13

vector field, 2
    complete, 83, 85, 93, 96, 104,
        110, 112, 156, 171
    semi-complete, 83, 91, 95, 97, 98,
        113, 156, 171, 174, 177, 180,
        182, 190, 192, 195
Visser–Ostrowski condition, 165, 186,
        192
Visser–Ostrowski quotient, 163, 165,
        186, 192
Vitali theorem, 6, 31

Wald's condition, ix, 163
Weierstrass convergence theorem, 6
Wolff's Lemma, 32, 33, 35, 54

List of Figures

0.1 The function $w = f(z)$. 2
0.2 The translation $f(z) = z + a$, $a = 4 + 2i$. 2
0.3 The rotation $f(z) = e^{i\pi\theta}z$, $\theta = -2\pi/3$. 3
0.4 The contraction $f(z) = kz$, $k = 1/3$. 3
0.5 The vector field $w = f(z)$. 3

1.1 An orthogonal circle to $\partial\Delta$ and its image under an automorphism. 14
1.2 A Möbius transformation of Δ_r. 16
1.3 A horocycle at the point $\zeta \in \partial\Delta$. 18
1.4 A nontangential approach region at a boundary point. 20
1.5 A Stolz angle at a boundary point. 21
1.6 Boundary behavior of a self-mapping of Δ. 22
1.7 Elliptic automorphism. 27
1.8 Hyperbolic automorphism. 28
1.9 Parabolic automorphism. 29
1.10 Uniqueness of a point ζ. 34

2.1 The points of anharmonic relation. 40

3.1 Boundary condition for $f \in \text{aut}(\Delta)$. 84
3.2 Boundary flow invariance condition. 86
3.3 Values of functions of Carathéodory's class. 90

4.1 Fractional iterations of the self-mapping $F(z)$. 110
4.2 The circle $\Gamma(z)$ and the sectors S and $\widetilde{S}$. 126
4.3 The nontangential convergence to τ. 130
4.4 The asymptotic behavior of the flow. 140

5.1 A starshaped domain ($0 \in \Omega$). 157
5.2 A starshaped domain ($0 \in \partial\Omega$). 158
5.3 The spiralshaped domain ($0 \in \partial\Omega$). 160

5.4   The set D.   168
5.5   Condition (A).   175
5.6   An approximation of a starshaped domain.   183

Erratum
ISBN 0-7923-7111-9

Revised titlepage

Semigroups in Geometrical Function Theory

by

David Shoikhet

Department of Mathematics
ORT Braude College
Karmiel, Israel
&
Department of Mathematics
Technion-Israel Institute of Technology
Haifa, Israel